Springer Theses

Recognizing Outstanding Ph.D. Research

Aims and Scope

The series "Springer Theses" brings together a selection of the very best Ph.D. theses from around the world and across the physical sciences. Nominated and endorsed by two recognized specialists, each published volume has been selected for its scientific excellence and the high impact of its contents for the pertinent field of research. For greater accessibility to non-specialists, the published versions include an extended introduction, as well as a foreword by the student's supervisor explaining the special relevance of the work for the field. As a whole, the series will provide a valuable resource both for newcomers to the research fields described, and for other scientists seeking detailed background information on special questions. Finally, it provides an accredited documentation of the valuable contributions made by today's younger generation of scientists.

Theses are accepted into the series by invited nomination only and must fulfill all of the following criteria

- They must be written in good English.
- The topic should fall within the confines of Chemistry, Physics, Earth Sciences, Engineering and related interdisciplinary fields such as Materials, Nanoscience, Chemical Engineering, Complex Systems and Biophysics.
- The work reported in the thesis must represent a significant scientific advance.
- If the thesis includes previously published material, permission to reproduce this must be gained from the respective copyright holder.
- They must have been examined and passed during the 12 months prior to nomination.
- Each thesis should include a foreword by the supervisor outlining the significance of its content.
- The theses should have a clearly defined structure including an introduction accessible to scientists not expert in that particular field.

More information about this series at http://www.springer.com/series/8790

Xiaolong Guo

Storage Policies and Maintenance Support Strategies in Warehousing Systems

Doctoral Thesis accepted by
the University of Science and Technology of China,
Hefei, People's Republic of China

Springer

Author
Dr. Xiaolong Guo
School of Management
University of Science and Technology of China
Hefei
People's Republic of China

Supervisor
Prof. Yugang Yu
School of Management
University of Science and Technology of China
Hefei
People's Republic of China

ISSN 2190-5053 ISSN 2190-5061 (electronic)
Springer Theses
ISBN 978-981-10-1447-5 ISBN 978-981-10-1448-2 (eBook)
DOI 10.1007/978-981-10-1448-2

Library of Congress Control Number:2016939576

Printed on acid-free paper

This Springer imprint is published by Springer Nature
The registered company is Springer Science+Business Media Singapore Pte Ltd.

The original version of the book frontmatter was revised: The author's affiliation has been corrected. The Erratum to this book frontmatter is available at 10.1007/978-981-10-1448-2_7

Parts of this thesis have been published in the following journal articles:

1. Yugang Yu, René de Koster, **Xiaolong Guo***. Class-based storage with a finite number of items: Using more classes is not always better. Production and Operations Management 24(8): 1235–1247, 2015. [Reproduced with permission from John Wiley and Sons]
2. **Xiaolong Guo**, Yugang Yu*, René de Koster. Impact of required storage space on storage policy performance in a unit-load warehouse. International Journal of Production Research 54(8): 2405–2418, 2016. [Reproduced with permission from Taylor & Francis]

*Corresponding author

To my father, my wife, and the memory of my mother

Supervisor's Foreword

The statement that "more classes will provide higher performance of warehousing systems under class-based storage policy" has been a general conclusion existed in the field of industrial and production engineering during the past several decades. Full turnover-based storage is claimed to provide the best system performance by creating different classes for each product. However, this well-known conclusion is rarely adopted in practice, and three- to five-class-based storage is usually implemented. This inconsistency has been an unsolved problem during the past 40 years, and no literature has theoretically demonstrated that too many classes degrade the system performance. This motivated us to investigate assumptions in the literature leading to this inconsistency. The cause of the inconsistency appears to be the assumption that "*the space needed per storage class equals the sum of the average inventory levels of the items in the class*". However, when items arrive in a batch quantity, sufficient space should be available to store all the items of the same class. Therefore, items sharing a storage class need more space than just their average stock level. If the number of storage classes increases, the number of items per class decreases and more space per item is needed on average, as the opportunity for space sharing decreases. This increases the average travel time for storing and retrieving items, and finally offsets the travel-time reduction, resulting from dividing items over a large number of classes according to their turnover. This trade-off was overlooked in the existing literature where most scholars assumed that the required storage space per item equals its average inventory level.

In addition to warehousing assignment policies, the maintenance support strategy for assembled equipment is another hot topic in the field of manufacturing and operations management. The performance-based contract (PBC) emerges as a new service model replacing the resource-based contract (RBC) in the after-market service supply chains. Existing studies indicate that system availability will be increased by 25–40 % under PBC than that under RBC. Most of previous studies aim to find the optimal contract terms leading to the minimal maintenance cost under a given availability target. However, enterprises in practice aim to obtain maximal profit instead of minimal maintenance cost. To the best of our knowledge, no study has yet discussed after-sales maintenance contract considering the profit-maximizing target from the viewpoint of the profitability of the system.

Taking into account the background of these two research topics, Dr. Guo started his Ph.D., expecting to find some answers in his thesis. First, relaxing of the popular assumption of conventional research. Dr. Guo studied the class-based storage policy considering a finite number of items in store. The results indicate that three- to five-class-based storage policy will be optimal for warehousing systems operated in a single-command mode. Second, PBC was studied with the consideration of the profitability of the maintained equipment. Results show that all customers will provide the same cost reimbursement for service supplier when the shared spare inventory is established. It also indicates that the maintenance service supplier can benefit from pooling the spare inventory only when the unit cost of its spare part is sufficiently low. His work provides us with important insights and suggestions for warehousing operations management and maintenance support management from both theoretical and practical perspectives.

Hefei
March 2016

Prof. Yugang Yu

Acknowledgments

First and foremost, I would like to express my deepest gratitude to my supervisor, Prof. Yugang Yu, for his patient guidance, continuous encouragement, and financial support during my Ph.D. His profound scientific attainments and rigorous attitude in academic research encouraged me to challenge the significant scientific problems. My ability to find and solve scientific problems extremely improved during the process. Professor Yu gave me the attentive care and guidance on the way of my growth in academic research, from proposing the research ideas to formulating the models, extracting the insights, and summarizing the contributions. It would have not been possible for me to complete the thesis without his boundless support. During my Ph.D., Prof. Yu encouraged me to present research results on academic conferences and to communicate with renowned scholars in related fields. Particularly, the visit to Rotterdam School of Management (RSM), Erasmus University from March to September 2013 is the most important academic experience for me, and it will be a precious fortune in my academic career. My attitude in research and skill of academic English are significantly improved during this visit. I would also like to take this opportunity to extend my gratitude to Prof. Liang Liang who provided me the first opportunity to start my academic career, and Dr. Liuyi Ling who guided me to accomplish my first research article.

My gratitude also goes to Prof. René de Koster who hosted and supervised me during my visit at RSM. His open-minded, enthusiastic, and rigorous attitude in academic research strengthened my determination to be a professional scholar. The help of Dr. Nima Zaerpour, Dr. Amir Hossein Gharehgozli, Dr. Dong Li, and other colleagues at RSM are very grateful.

I am thankful for the sufficient encouragement and selfless help from professors and researchers from different countries and universities. There are many of them to name for appreciation. They are Prof. Zhimin Huang from Adelphi University, Prof. Jinfeng Yue from Middle Tennessee State University, and Dr. Sean Zhou from the Chinese University of Hong Kong. Colleagues of the School of Management, University of Science and Technology of China including Prof. Feng Yang, Prof. Gongbing Bi, Dr. Shaofu Du, Dr. Tengfei Nie, and members of

Prof. Yu's research team, including Dr. Zhaofu Hong, Dr. Ye Shi, Ruiqi Hou, Huan Yu, Hu Yu, Xiaoya Han, Wan Wu, Yu Xue, and many others helped me quite a lot during my Ph.D.

Many thanks to my families for their endless support, inspiration, and love. Especially my father, who raised me and my younger sister alone after my mother passed away. His endless support is the spiritual source for me to accomplish my Ph.D. Special thanks to my late mother for raising me, and this work is dedicated to the memory of her. I thank my wife, Chenchen Yang, for her love and immense support throughout my Ph.D. journey. Without her understanding and support, it would have not been possible for me to complete this thesis.

Publication of this book was financially supported by the National Natural Science Foundation of China (Nos. 71225002, 71501174), and the China Postdoctoral Science Foundation (No. 2014M560523). Editor Wayne Hu is acknowledged for his help in the publication of this work.

Contents

Acronyms

AS/R	Automated storage and retrieval
AS/RS	Automated storage/retrieval system
BM	Basic travel-time model
cdf	Cumulative distribution function
CM	Conventional travel-time model
CTM	Continuous travel model
DSM	Discrete-stochastic model
EOQ	Economic order quantity
NSIT	Non-square-in-time
PBC	Performance-based contracts
pdf	Probability density function
RSS	Required storage space
SIT	Square-in-time
SM	Solution methodology
SMB	Solution methodology with Benders ABC curve
S/R	Storage/retrieval
TDM	Travel distance model
TEUs	Twenty-foot equivalent units

List of Figures

List of Tables

Chapter 1
Introduction

Abstract This chapter provides the introduction of this book. Section 1.1 gives the basic research background and significance of the body. Section 1.2 presents a brief review of the related literature. Section 1.3 discusses the research contents and research methodologies used in this book. Section 1.4 provides the outline of the book.

1.1 Research Background and Significance

Logistics management has become one of the most popular topics in both economic development and scientific research since the end of the last century because of the rapid development of the Internet and electronic commerce. As critical nodes in supply chains, the order response efficiency of warehousing systems have become an important indicator of modern logistics management. Consequently, improving operational efficiency has become a research hotspot in the industrial engineering field, wherein the optimization of class-based storage policy in warehousing systems receives great attention from scholars. In addition, the maintenance support strategy of assembled equipment in the after-sales market also attracts considerable attention from researchers in the environment of that the economic system is transforming toward a service-based economy. Based on this business background, the current book examines the most important issues related to the order response efficiency of warehousing systems: the storage policies for items stored in the system and the maintenance support of storage/retrieval (S/R) machines.

1.1.1 Class-Based Storage

Class-based storage policy has received considerable attention along with the popularization of automated storage/retrieval systems (AS/RS) since the 1960s. Heskett (1963, 1964) proposed the famous cube-per-order index rule, which is also known as the full turnover storage policy. This rule is a method used to maximize the oper-

X. Guo, *Storage Policies and Maintenance Support Strategies in Warehousing Systems*, Springer Theses, DOI 10.1007/978-981-10-1448-2_1

ational efficiency of AS/RS and a benchmark used to evaluate other storage policies (e.g., class-based storage policy). Thereafter, following these papers and the study of Hausman et al. (1976) on two and three class-based storages, many research articles have been published on the topic of class-based storage policy (Ang et al. 2012; Koster et al. 2007; Eynan and Rosenblatt 1994; Rosenblatt and Eynan 1989; Teunter et al. 2010; Yu and Koster 2009). Although most of these studies have focused on the optimal storage assignment policy for a given number of classes, none of them have considered the optimization of the number of classes for warehousing systems. As Kampen et al. (2012) concluded in a review paper, no guideline is available to determine the optimal number of classes (for class-based storage policy), and "the best performance could theoretically be expected to be achieved by creating different classes for each product". However, this theoretically perfect storage policy is rarely used in reality. On the contrary, class-based storage policies with three to five classes is adopted widely.

No existing study has been conducted on the reason for the rare adoption of a full turnover-based storage policy in practice as all studies think that full turnover-based storage policy will provide the best performance for a system with class-based storage policy (Kampen et al. 2012). In investigating the existing literature, we find that most publications implicitly or explicitly assume that the total required storage space does not depend on the number of classes in modeling and is equal to the average storage space of the item. On the one hand, this assumption is valid only when the number of items in each class is sufficiently large (infinite). Because multiple items are stored randomly and share a common storage space within each storage class. Items are replenished to the system at different points in time, and any available empty storage location in the class can be used for storing it when an item is replenished. On the other hand, when a full turnover-based storage policy applied, only one item exists in each individual class. This situation is contradictory to the condition made for the abovementioned assumption to be valid. This contradiction motives us to investigate whether full turnover-based storage can provide optimal performance. Consequently, Part I of this book revisits the optimization of a class-based storage policy by relaxing the assumption "the required storage space of all the items equals their average inventory level" and by considering a finite number of items in a warehousing system. The results fill the theoretical gap and provide useful suggestions for warehouse managers.

1.1.2 Maintenance Support Strategies

In achieving the high performance of warehousing operations in the e-commerce era, maintenance support and availability management for warehousing equipment are critical strategies aside from item assignment policies. Maintenance support for assembled equipment in the after-sales market has received considerable attention from scholars in the field, thus indicating that the economic system is transforming toward a service-based economy. In the after-sales, service supply chains consist of customers (i.e., the enterprises with the equipment) and suppliers (i.e., the companies providing maintenance services), and service contracts are usually designed according to the minimum maintenance cost under an availability requirement. Resource-based contracts and performance-based contracts are widely discussed in the existing literature as two major types of contracts used to describe the relationship between customers and maintenance suppliers (Kim et al. 2007, 2010; Oliva and Kallenberg 2003; Roels et al. 2010). Kim et al. (2007) examined the performance-based contracting model through a sequential game to minimize the maintenance cost of customers. While for mission-critical systems with infrequent restoration and recovery, the supplier bears the expensive cost of committing the necessary spare inventory because the failures occur infrequently and the inventory will be idle most of time. To close this gap, Kim et al. (2010) introduced a performance-contracting model for the infrequent restoration and recovery of mission-critical systems. Thereafter, the interaction between the component reliability and spare inventory decisions when performance-based contracts are applied is studied (Kim et al. 2011).

Most of the existing studies on performance-based contracts aim to minimize the maintenance cost according to a given availability requirement. However, the objective of enterprises in practice is not to maintain the equipment at given availability targets but to create profit. In other words, the objective of customers should be the maximum profit instead of the minimum maintenance cost. Thus, on the background of operations management of warehousing systems, Part II of this book examines the maintenance support strategies of assembled systems subject to the optimal profit regarding the profitability of the systems in the market. The findings will enrich theory of performance-based contracts and maintenance management in the after-sales service supply chains and provide decision support for enterprises in the competitive economic environment.

1.2 Related Literature

Class-based storage is widely discussed in many operations management textbooks and scientific articles (for example Eynan and Rosenblatt 1994; Johnson and Brandeau 1996; Kouvelis and Papanicolaou 1995; Rosenblatt and Eynan 1989). Different from the famous cube-per-order index rule (also known as full turnover storage) (Heskett 1963, 1964; Koster et al. 2007), class-based storage divides items into different classes according to the ABC demand curve of the warehousing system. The first research on class-based storage is Hausman et al. (1976), which examined the optimal assignment policy for two and three class-based storage policies. On the basis of this research, Rosenblatt and Eynan (1989) and Eynan and Rosenblatt (1994) formulated a travel-time model for n-class-based storage and addressed the benefits of class-based storage by increasing the number of classes. Most research on class-based storage implicitly or explicitly assumes that the total required storage space is equal to the average storage space of the item, which is valid when the number of items in every class is sufficiently large. Within each storage class, multiple items replenished to the system at different points in time are stored randomly and they share a common storage space. When a replenishment order of an item arrives, any available empty storage location in its class can be used for storing it. As a result, if the number of items in a class is infinite, then the required storage space of the class will approximately equal the average total inventory level of all items in the class. However, this condition cannot be invalid in practice especially for full turnover-based storage. To investigate the influence of the finiteness of items in each class, this book will revisit the typical travel-time models by considering a finite number of items.

Maintenance contracts between customers and suppliers in after-sales service supply chains have also received significant attention from academic scholars. The two contract types discussed in the existing literature are resource-based contracts and performance-based contracts (Kim et al. 2007, 2010; Roels et al. 2010). In the literature, performance-based contracts emerge as a new service model replacing resource-based contracts in asset-intensive industries. Kim et al. (2007) examined performance contracting to minimize the maintenance cost for customers with a given availability requirement for its systems. Customers pay a part of suppliers' maintenance cost, which can be reduced by suppliers' cost reduction effort as compensation, and penalize the suppliers depending on the number of backorders of each subsystem. Thereafter, the authors extended the work from two directions by focusing on the interaction between the component reliability and the spare inventory decisions (Kim et al. 2011), and a two-stage sequential game model that consists of multiple risk-neutral suppliers and a customer for new products (Kim 2011). These three studies complement one another and provide significant managerial insights into customers and suppliers in performance-based contracting scenarios. In addition, Mirzahosseinian and Piplani (2011) found that increasing the components' reliability and efficiency of repair facility are two more effective measures than setting a high spare inventory level. Guajardo et al. (2012) empirically investigated the relationship between product reliability and after-sales contract based on a proprietary data set provided by Rolls-Royce (a major manufacturer of aircraft engines). Their estimation indicates that product availability under performance-based contracts increases by 25–40 % compared with that under resource-based contracts. However, all of the mentioned publications only consider the minimum maintenance cost with given availability targets, and none of them considers profit optimization as the objective, which is the actual concern of the customers in reality. Therefore, Part II of this book investigates the problem of optimizing the profit of customers by relaxing the constraint of availability targets under performance-based contracts.

1.3 Research Contents and Methodologies

On the basis of the research background and significance and the key studies reviewed in the previous sections, this section introduces the research contents and major methodologies.

1.3.1 Research Contents

The research contents of this book are divided into two parts: optimizing the storage policies in warehousing systems with a finite number of items and optimal maintenance support strategies for warehousing equipment under performance-based contracts.

(I) The optimization of class-based storage policy for warehousing systems with finite number of items is studied by relaxing the common assumption in conventional research, that is, the required storage space equals its average inventory level. A basic travel-time model is built for an AS/RS with a continuous square-in-time (SIT) storage rack and is extended to different storage racks including non-SIT (NSIT) racks and discrete racks. Different item demand structures include various demand curves and stochastic demand information. In addition, the performance of storage policies is evaluated in a unit-load warehouse with parallel storage aisles.

(II) Maintenance support strategies subject to the maximum expected profit considering system profitability in the market under performance-based contracts are studied in an after-sales service supply chain through a sequential game with multiple leaders and multiple followers. Integrated (i.e., self-maintenance scenario) and decentralized games are analyzed. Optimal decisions on contract terms (of customers), spare inventories (of service suppliers), and cost reduction efforts (of service suppliers) are investigated.

1.3.2 Research Methodologies

The two major research methodologies employed in this book are dynamic programming and game theory.

(1) Dynamic programming is widely used in solving problems in production scheduling (Florian et al. 1980), loading (Hodgson 1982), sequencing (Psaraftis 1980), inventory management (Veinott 1965), equipment replacement (Hastings 1968), and production optimization (Findlay et al. 1989) in the fields of industrial engineering, enterprise management, operations management, military management, and industrial and agricultural production (Bellman 1956). The objective stated in the first part of this book is to find the optimal item assignment policy that minimizes the average one-way travel-time for warehousing systems. Dynamic programming is used to solve the travel-time models of this part.

(2) Game theory is widely used in modeling the interactions of decision makers and in deducing their game equilibrium in the fields of economics and management (Von Neumann and Morgenstern 2007). In the literature, Nash game and Stackelberg sequential game are two of the most popular game models (Fraja

and Delbono 1990). In the second part of this book, game theory is adopted to model the decision interactions in the after-sales service supply chain. In particular, Stackelberg sequential game is used to describe the relationship between customers and service providers, and Nash game is adopted to describe the relationships among multiple customers.

1.4 Outline of the Book

As shown in Fig. 1.1, this book consists of six chapters arranged in the following manner:

This chapter gives the introduction of this book by providing the basic research background and significance, a brief review of key studies, and the research contents and methodologies.

Part I of the book consists of two chapters. Chapter 2 examines the optimal storage assignment policy under class-based storage in an AS/RS. A basic travel-time model with a continuous SIT storage rack and economic ordering quantity (EOQ) replacement is proposed through the consideration of a finite number of items stored. Thereafter, extended models are presented by relaxing related assumptions, such as discrete storage rack, NSIT rack, stochastic item demand, and Bender's ABC demand curve.

Chapter 3 analyzes the performance of storage policies (i.e., random storage, class-based storage, and full turnover-based storage) in a unit-load warehouse with parallel storage aisles. A discrete travel-time model is built in consideration of the realistic required storage space of the warehouse. Optimal item classifications are obtained for the warehouse with different item demand structures.

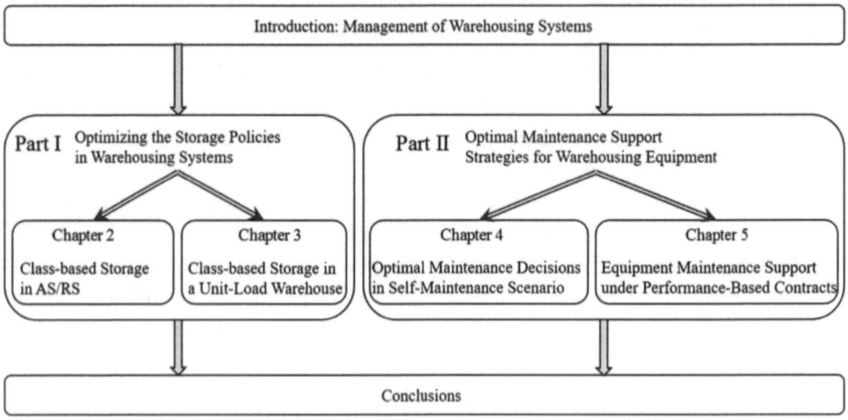

Fig. 1.1 Structure of the book

Part II of the book consists of Chaps. 4 and 5. Chapter 4 investigates the relationship between the availability requirement of assembled systems and its corresponding profitability in the market. Optimal maintenance decisions are obtained for a storage enterprise when it perform maintenance support on its own. Furthermore, an alliance model with multiple enterprises is analyzed to verify the polling effect when a shared spare inventory is available.

Chapter 5 examines the optimal maintenance support strategies for customers in an after-sales service supply chain under performance-based contracts. The interactions between customers and suppliers are modeled as a Stackelberg sequential game, and competition among multiple customers is described as a Nash game. Optimal decisions on spare inventories and cost reduction efforts are deduced for suppliers with given contract terms, and optimal contract terms for customers based on the responses of the suppliers are also provided.

Finally, Chap. 6 concludes the book by summarizing the contributions and insights, and presenting the necessary limitations and future research directions.

References

Ang M, Lim YF, Sim M (2012) Robust storage assignment in unit-load warehouses. Manag Sci 58(11):2114–2130

Bellman R (1956) Dynamic programming and lagrange multipliers. Proc Natl Acad Sci U.S.A. 42(10):767

De Koster R, Le-Duc T, Roodbergen KJ (2007) Design and control of warehouse order picking: a literature review. Eur J Op Res 182(2):481–501

Eynan A, Rosenblatt MJ (1994) Establishing zones in single-command class-based rectangular as/rs. IIE Trans 26(1):38–46

Findlay P, Kobbacy K, Goodman D (1989) Optimization of the daily production rates for an offshore oilfield. J Op Res Soc. pp 1079–1088

Florian M, Lenstra JK, Rinnooy Kan A (1980) Deterministic production planning: algorithms and complexity. Manag sci 26(7):669–679

Fraja G, Delbono F (1990) Game theoretic models of mixed oligopoly. J Econ Surv 4(1):1–17

Guajardo JA, Cohen MA, Kim SH, Netessine S (2012) Impact of performance-based contracting on product reliability: an empirical analysis. Manag Sci 58(5):961–979

Hastings N (1968) Some notes on dynamic programming and replacement. J Op Res Soc. pp 453–464

Hausman WH, Schwarz LB, Graves SC (1976) Optimal storage assignment in automatic warehousing systems. Manag Sci 22(6):629–638

Heskett JL (1963) Cube-per-order index-a key to warehouse stock location. Transp Distrib Manag 3(1):27–31

Heskett JL (1964) Putting the cube-per-order index to work in warehouse layout. Transp Distrib Manag 4(8):23–30

Hodgson TJ (1982) A combined approach to the pallet loading problem. IIE Trans 14(3):175–182

Johnson ME, Brandeau ML (1996) Stochastic modeling for automated material handling system design and control. Transp Sci 30(4):330–350

Kim SH (2011) Strategic reliability investments in multi-indenture supply chains. Tech. rep., Working paper, Yale University

Kim SH, Cohen MA, Netessine S (2007) Performance contracting in after-sales service supply chains. Manag Sci 53(12):1843–1858

Kim SH, Cohen MA, Netessine S, Veeraraghavan S (2010) Contracting for infrequent restoration and recovery of mission-critical systems. Manag Sci 56(9):1551–1567

Kim SH, Cohen MA, Netessine S (2011) Reliability or inventory? analysis of product support contracts in the defense industry

Kouvelis P, Papanicolaou V (1995) Expected travel time and optimal boundary formulas for a two-class-based automated storage/retrieval system. Int J Prod Res 33(10):2889–2905

Mirzahosseinian H, Piplani R (2011) A study of repairable parts inventory system operating under performance-based contract. Eur J Op Res 214(2):256–261

Oliva R, Kallenberg R (2003) Managing the transition from products to services. Int J Ser Ind Manag 14(2):160–172

Psaraftis HN (1980) A dynamic programming approach for sequencing groups of identical jobs. Op Res 28(6):1347–1359

Roels G, Karmarkar US, Carr S (2010) Contracting for collaborative services. Manag Sci 56(5):849–863

Rosenblatt MJ, Eynan A (1989) Note deriving the optimal boundaries for class-based automatic storage/retrieval systems. Manag Sci 35(12):1519–1524

Teunter RH, Babai MZ, Syntetos AA (2010) Abc classification: service levels and inventory costs. Prod Op Manag 19(3):343–352

Van Kampen TJ, Akkerman R, Pieter van Donk D (2012) Sku classification: a literature review and conceptual framework. Int J Op Prod Manag 32(7):850–876

Veinott AF Jr (1965) Optimal policy for a multi-product, dynamic, nonstationary inventory problem. Manag Sci 12(3):206–222

Von Neumann J, Morgenstern O (2007) Theory of games and economic behavior. Princeton university press, Princeton

Yu Y, de Koster RB (2009) Optimal zone boundaries for two-class-based compact three-dimensional automated storage and retrieval systems. IIE Trans 41(3):194–208

Part I
Optimizing the Storage Policies of Warehousing Systems

The first part of this book focuses on the storage policies of warehousing systems and analyzes the ways of improving operational efficiency based on different storage policies. The discussions on this part are found in Chaps. 2 and 3. In Chap. 2, by considering a finite number of items stored in the system, we relax the following basic assumption in the literature: the required storage space for all the items equals their average inventory level, which is valid only if an infinite number of items are stored in each storage region. First, we provide the basic one-way travel-time model based on a SIT continuous storage rack. Second, we relax the assumptions by considering NSIT storage racks, different types of ABC demand curve, different numbers of stored items, discrete storage racks, and stochastic item demands to validate the robustness of the basic model. Thereafter, in Chap. 3, we examine the performance of different storage policies (i.e., random storage, class-based storage, and turnover-based storage) in a unit-load traditional warehouse with a finite number of items by considering the effect of a realistic required storage space.

Chapter 2
Class-Based Storage with a Finite Number of Items in AS/RS

Abstract Class-based storage is widely studied in the literature and applied in practice. It divides all stored items into a number of classes according to their turnover. A class of items with a high turnover is allocated to a region close to the warehouse depot. Studies have shown that the use of more storage classes leads to a shorter travel time for storing and retrieving items. A basic assumption in this literature is that the required storage space for all the items is equal to their average inventory level, which is valid only if an infinite number of items are stored in each storage region. Therefore, this chapter revisits class-based storage by considering a finite number of items and by relaxing the assumption that the "required storage space of all the items equals their average inventory level". We develop a travel-time model and algorithm that can be used for determining the optimal number and the boundaries of storage classes in warehouses. Different from the results of conventional research, our findings illustrate that a small number of classes is generally optimal. In addition, we find that travel time is fairly insensitive to the number of storage classes in a wide range around the optimum. This finding suggests that managers can select a near-optimal number of storage classes in an easy way, and they should not be worried about the effect of storage-class reconfigurations. We validate our findings for various cases, including different ABC demand curves, space-sharing factors, number of items, storage rack shapes, discrete storage locations, and stochastic item demand.

2.1 Research Background

Class-based storage is the most commonly used storage policy in practice and is widely discussed in many operations management textbooks (Tompkins et al. 2010; Heragu 2006; Adams 1996) and scientific papers (Rosenblatt and Eynan 1989; Kouvelis and Papanicolaou 1995; Johnson and Brandeau 1996; Gu et al. 2007; Eynan and Rosenblatt 1994; De Koster et al. 2007). It divides stored items into different classes (using three classes is common in practice) according to the ABC demand curve (see Fig. 2.1). In case of ABC class-based storage, a relatively small number of highly demanded items are grouped as A-class items and are then stored in a

X. Guo, *Storage Policies and Maintenance Support Strategies in Warehousing Systems*, Springer Theses, DOI 10.1007/978-981-10-1448-2_2

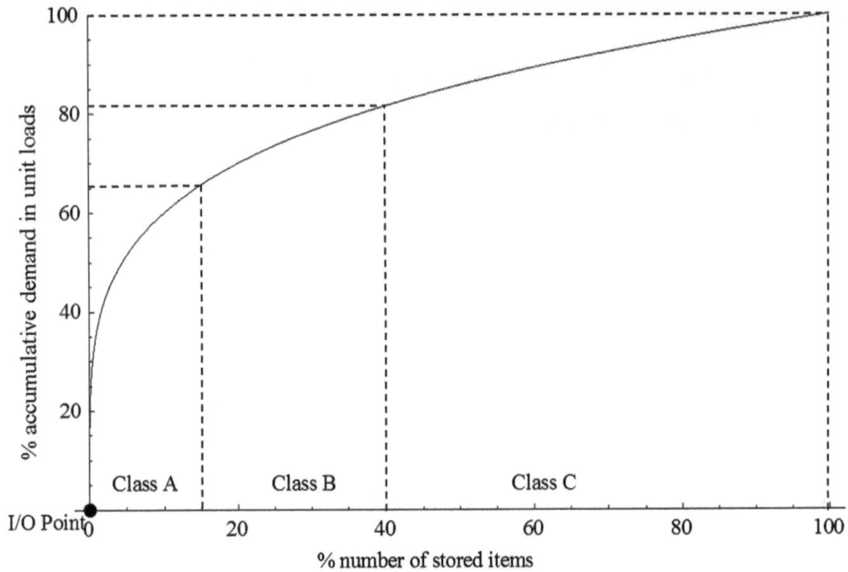

Fig. 2.1 An example of an ABC demand curve

warehouse region closest to the depot (the entrance and exit position). Grouped as C-class items, rarely demanded items are stored in the region farthest from the depot. Within each class, items are stored randomly.

Figure 2.2 illustrates the side view of a SIT storage rack with an example of an ABC class-based storage as used in automated storage and retrieval (AS/R) warehousing systems. In such a system, the optimal boundary of each region is square because the storage/retrieval (S/R) machines can drive and lift simultaneously. This capability leads to a Chebyshev distance metric (Bozer et al. 1990) used to measure the distance between a storage location and the depot.

Hausman et al. (1976) modeled and analyzed the two- and three-class-based storage policies; Rosenblatt and Eynan (1989) and Eynan and Rosenblatt (1994) formulated a travel-time model for n-class-based storage and addressed the benefits of class-based storage by increasing the number of classes. Following these studies and the earlier paper, most studies on class-based storage (Eynan and Rosenblatt 1994; Larson et al. 1997; Gu et al. 2007; Yu and Koster 2009) have implicitly or explicitly assumed that the total required storage space does not depend on the number of classes in modeling. This assumption is valid when the number of items in each class is sufficiently large (infinite). Within each storage class, multiple items are stored randomly and share a common storage space. They are replenished in the system at different points in time. When an item is replenished, any available empty storage location in its class can be used for storing it. As a result, if the number of

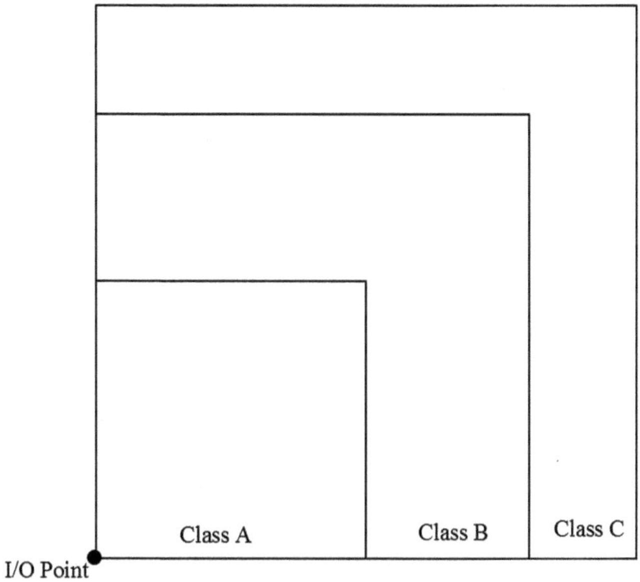

Fig. 2.2 Side view of a rack with ABC class-based storage regions

items in a class is infinite, then the required storage space of the class is approximately equal to the average total inventory level of all items in the class. With this assumption, conventional research has indicated that an increase in the number of classes reduces the average travel time for storing or retrieving items (see the curve indicating conventional research in Fig. 2.3).

However, the abovementioned widely cited finding is inconsistent with the practice in which only a few (three to five) classes are usually implemented (Roodbergen and Vis 2009). To the best of our knowledge, no study has theoretically demonstrated that an excessive number of classes degrade system performance. This deficiency motivates us to investigate all assumptions made in the literature. The cause of the inconsistency appears to be the assumption that regardless of the number of storage classes, the space needed for each storage class is equal to the sum of the average inventory levels of the items in the class. This assumption can be justified in the case of an infinite number of items stored for each product class. However, every time an item is received, sufficient space should be available for storing the entire batch. Therefore, items sharing a storage class need more space than just their average stock level. If the number of storage classes increases, then the number of items in each class decreases, and more space is needed for each item as the opportunity for space-sharing decreases. This condition increases the average travel time for storing and retrieving items and finally offsets the travel-time reduction resulting from dividing items over a large number of classes according to their turnover. This tradeoff has

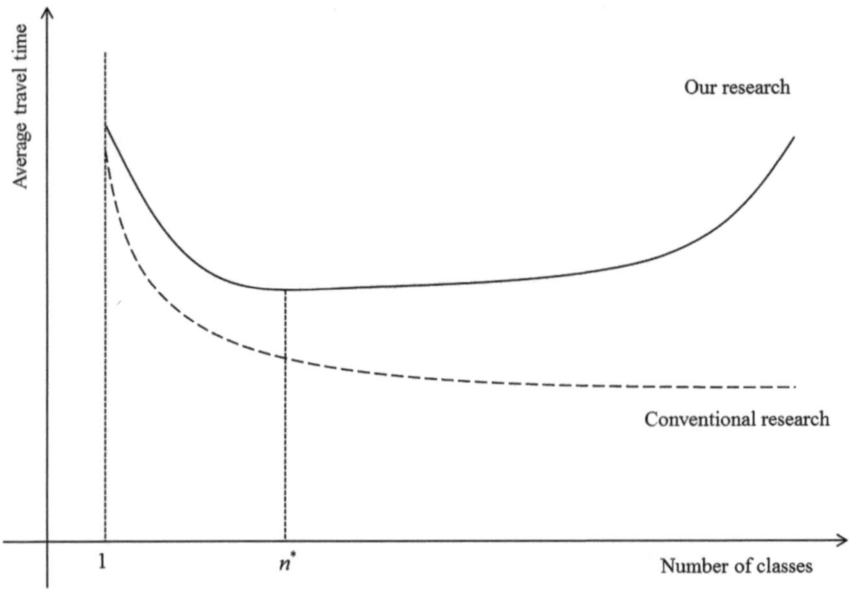

Fig. 2.3 Travel times in two different lines of research

not been investigated in the literature. Relaxing other common assumptions, such as deterministic demand, continuous and SIT racks, and use of the classic EOQ replenishment policy, does not affect the existence of this tradeoff. Therefore, we first relax the (implicit) finite number of items assumption and then validate the robustness of our findings by relaxing other assumptions made in the literature.

To investigate this tradeoff overlooked in the existing literature and to deduce managerial insights for warehouse managers, this chapter develops a new expression for estimating the required storage space as a function of the number of storage classes. Based on this expression, a travel-time model is developed from which the optimal number of classes and their boundaries can be determined for a warehouse with a finite number of items. The model is solved using dynamic programming with time complexity $O(N^3)$, where N is the number of stored items in the system. The results demonstrate that travel time is commonly a bowl-shaped function of the number of classes as shown in Fig. 2.3. This flat curve shape yields some important findings. First, beyond a small number of classes, an increase in the number of classes cannot reduce travel time. Second, a small number of classes is usually enough to yield an optimal solution. Third, travel time appears to be insensitive to the number of classes in a wide range around the optimum (corresponding to the bottom of the bowl-shaped curve in Fig. 2.3). This finding implies that warehouse managers should not hesitate to change their number of classes if necessary because "they cannot go wrong".

2.2 Problem Description and Conventional Travel-Time Model

This section describes the system studied and develops the mathematical model. The traditional model in the literature is also revisited.

2.2.1 *Problem Description*

Without loss of generality, the basic idea of n-class-based storage is usually discussed in the abovementioned literature with a basic automated warehousing system: the AS/RS. This system consists of an S/R machine, a continuous storage rack, and one depot where all items enter and leave the system. Items can be finished goods, work-in-process, or raw materials, which are stored on standardized unit loads (e.g., pallets or totes) in the AS/RS. The system works as follows: when a storage unit load arrives at the depot of the system, the machine retrieves and transports it to any given storage location in the rack. When a stored unit load is requested, the machine picks it up and moves it to the depot. The system has the following properties:

1. All storage locations are the same size for storing standardized unit loads.
2. The depot is located on the lower-left side of the storage rack.
3. The continuous-space storage rack is SIT; the time for the machine to move from the depot to the most distant column is equal to the time for the machine to move from the depot to the most distant tier. The machine travels in horizontal and vertical directions simultaneously, thus resulting in a Chebyshev distance metric used to measure the distance between a storage location and the depot Roodbergen and Vis (2009). An extended model incorporating NSIT racks is given in Sect. 2.4.1. A discrete space rack is modeled in Sect. 2.4.3.
4. The capacity of the machine is one unit load. The machine operates in a single-command mode, and it stores or retrieves one unit load each time.
5. The pick-up or deposit time for the machine to load or unload a unit load is constant and ignored.
6. The turnover of each item is measured as the number of unit loads requested in a unit-time period, such as a week, a month, or a year, and is determined by the ABC demand curve given in Eq. (2.1). All the items are ranked according to their marginal contribution to the total turnover; an item that has a smaller contribution is indexed with a larger number. We extend this method to the stochastic demand in Sect. 2.4.3.
7. Item inventories are replenished according to the classic EOQ model.

As used in conventional research, the ABC demand curve is a plot of ranked cumulative percentage expected demand per unit time, $G(i)$, and is modeled by the following:

$$G(i) = i^s = \int_0^i D(j)dj \bigg/ \int_0^1 D(j)dj, \qquad 0 < s \le 1, \tag{2.1}$$

where i is the item at the ith percentile in the ranked sequence of all items, $D(i)$ is the demand of item i per unit time, and s is the shape factor of the ABC demand curve. Given $s = 0.222$, we know that 20 % of the total items (i.e., $i = 20\,\%$) contribute $G(i) = i^s = (20\,\%)^{0.222} = 70\,\%$ of the total demand. A lower s means a more skewed ABC demand curve. For example, $s = 0.222$ represents a 20 %/70 % demand curve that is more skewed than a 20 %/50 % ABC demand curve for $s = 0.431$. We relax the ABC curve function with another demand curve function in Sect. 2.4.2. By normalizing the total demand $\int_0^1 D(j)dj = 1$, we obtain the following without loss of generality:

$$D(i) = dG(i)/di = si^{s-1}, \qquad 0 < s \le 1, \tag{2.2}$$

according to Hausman et al. (1976) and Rosenblatt and Eynan (1989).

Given the abovementioned system properties and the item demands determined by Eq. (2.2), we intend to find the average one-way travel time for storing or retrieving a unit load in a class-based storage system. The one-way travel time is the travel time from the depot to a unit-load storage location.

The class-based storage policy divides the storage space into n regions. Region k is dedicated to storing items of class k, $k = 1, 2, \ldots, n$. As shown in Fig. 2.1, a region with items of high demand is located close to the depot. Items are randomly stored in each region. Furthermore, the regions are L-shaped in a SIT storage system.

Based on the description given so far, the notations used in this chapter are given and defined in Table 2.1.

With the abovementioned notations and according to Rosenblatt and Eynan (1989), the average one-way travel time (called "travel time" hereafter) in an n-class system, T_n, can be formulated as the follows:

$$T_n = \frac{\sum_{k=1}^n t_k \Lambda(k)}{\sum_{k=1}^n \Lambda(k)} = \sum_{k=1}^n t_k \left(\frac{\Lambda(k)}{\sum_{k=1}^n \Lambda(k)} \right), \tag{2.3}$$

where $\Lambda(k) \Big/ \sum_{k=1}^n \Lambda(k)$ is the weighted retrieval rate of class k, and $\sum_{k=1}^n \Lambda(k)$ is the total turnover in the whole system. Rosenblatt and Eynan (1989) simplified the calculation using Eq. (2.2) to obtain the following:

Table 2.1 Notations used in this chapter

Notations	Definitions
i	Index of the ith item. An item with a lower demand has a larger index
j	Index of the jth storage location (or unit load). A location closer to the depot has a smaller index
n	Number of classes in the storage system
k	Index of the kth class, $k = 1, 2, \ldots, n$
i_k	Index of the item with the lowest turnover in class k
j_k	Storage location (a corresponding unit load) farthest from the depot in class k. It also corresponds to the total required storage space of items 1 to k
t_k	Average one-way travel time for storing/retrieving a unit load of class k
R_k	One-way travel time for storing/retrieving a unit load at the further boundary of class k
G_k	100×(cumulative demand for the first k classes)/(the total demand of all items in a unit-time period)
$\Lambda(k)$	Total turnover, in number of unit loads per unit-time period of all items stored in class k
T_n	Average one-way travel time of a unit load for an n-class storage system

$$\Lambda(k) \Big/ \sum_{k=1}^{n} \Lambda(k) = G_k - G_{k-1} = i_k^s - i_{k-1}^s, \qquad k = 1, 2, \ldots, n. \tag{2.4}$$

Furthermore, in case of a SIT system, according to Hausman et al. (1976) and Rosenblatt and Eynan (1989), t_k can be obtained as the follows:

$$t_k = \frac{2\left(R_k^3 - R_{k-1}^3\right)}{3\left(R_k^2 - R_{k-1}^2\right)}, \qquad k = 1, 2, \ldots, n. \tag{2.5}$$

Consequently, by substituting Eqs. (2.4) and (2.5) into Eq. (2.3), the average travel time for the system can be rewritten as follows:

$$T_n = \sum_{k=1}^{n} \frac{2\left(R_k^3 - R_{k-1}^3\right)}{3\left(R_k^2 - R_{k-1}^2\right)} \left(i_k^s - i_{k-1}^s\right). \tag{2.6}$$

To minimize T_n in Eq. (2.6), we must derive the relationship between R_k and i_k (or G_k), $k = 1, 2, \ldots, n$. When the relationship is obtained, travel time can be minimized through the optimization of either R_k or i_k, for $k = 1, 2, \ldots, n$.

In the next subsection, the conventional travel-time model based on the assumption that "the required storage space of all the items equals their average inventory level" is revisited.

2.2.2 Conventional Travel-Time Model

Conventional research (Rosenblatt and Eynan 1989; Hausman et al. 1976; Eynan and Rosenblatt 1994) assumes that the total required storage space for storing all items is equal to the total average inventory level of the items regardless of the number of classes of the system

$$A = L = \int_0^1 Q(i)/2di = \sqrt{2Ks}\Big/(s+1), \tag{2.7}$$

where A is the total required storage space (in number of unit load locations) for storing all items, L is the total average inventory level of the items, $Q(i)$ is the economic order quantity of item i, and K is the ratio of order cost to holding cost and is assumed to be equal for all items. On the basis of the abovementioned assumptions, Rosenblatt and Eynan (1989) provided the relationship between R_k and G_k as $G_k = R_k^{4s/(s+1)}$ with $G_k = i_k^s$. As a result, the conventional model (hereafter called as Model CM) can be defined as follows:

Model CM:

$$\begin{aligned}
&Min \quad T_n = \sum_{k=1}^{n} \frac{2\left(R_k^3 - R_{k-1}^3\right)}{3\left(R_k^2 - R_{k-1}^2\right)}\left(R_k^{4s/(s+1)} - R_{k-1}^{4s/(s+1)}\right),\\
&s.t. \quad 0 = R_0 < R_{k-1} < R_k < R_n = 1,\\
&d.v. \quad R_k, \quad k = 1, 2, \ldots, n-1.
\end{aligned} \tag{2.8}$$

Model CM can be solved using the following recursive equation proposed by Rosenblatt and Eynan (1989):

$$T_n = \frac{2\left(1 - R_{n-1}^3\right)}{3\left(1 - R_{n-1}^2\right)}\left(1 - R_{n-1}^{4s/(s+1)}\right) + R_{n-1}^{(5s+1)/(s+1)} T_{n-1}. \tag{2.9}$$

2.3 Travel-Time Model with a Finite Number of Items

This section relaxes the assumption of conventional research by adopting a finite number of items in the system. The total required storage space is not simply equal to the average inventory level similar to the Model CM but becomes a function of the number of classes and the number of items in each class. In Sect. 2.3.1, the required storage space for each storage class and the relationship between R_k and i_k are derived. Section 2.3.2 presents our basic model that considers the required storage space. A solution methodology for the model is provided in Sect. 2.3.3.

2.3.1 Required Storage Space Function and Relationship Between R_k and i_k

If the number of items sharing a common storage space is finite, then the required storage space of an item depends on several factors such as the number of items sharing the space, the skewness of the ABC demand curve (s), the inventory replenishment policies, and the ratio of order cost to holding cost (K). We first determine the required storage space of an item as a function of the number of items in the same shared space by considering the replenishment quantity $Q(i)$ that incorporates the other factors.

The function $a_i(N_k)$ denotes the space required (average over time) to store item i in class k together with $(N_k - 1)$ other items, where N_k represents the number of items sharing a common storage space within class k. A large number of storages and retrievals are simulated for obtaining the presentation of $a_i(N_k)$, and the general shape is sketched in Fig. 2.4. In particular, when $N_k = 1$, the storage method turns into a dedicated storage, and the required storage space for item i is equal to its order quantity $Q(i)$. When $N_k = +\infty$, the method turns into the situation with an infinite number of items, and the required storage space for item i is now equal to its average inventory level, $Q(i)/2$ (Hausman et al. 1976; Rosenblatt and Eynan 1989). For $1 < N_k < +\infty$, the value of $a_i(N_k)$ is between $Q(i)$ and $Q(i)/2$, and it decreases in N_k convexly as shown in Fig. 2.4. The relationship can be represented by the following mathematical presentation:

$$a_i(N_k) = 0.5\left(1 + N_k^{-\varepsilon}\right) Q(i) = 0.5\left(1 + N_k^{-\varepsilon}\right)\sqrt{2KD(i)}, \tag{2.10}$$

where ε is the space-sharing factor.

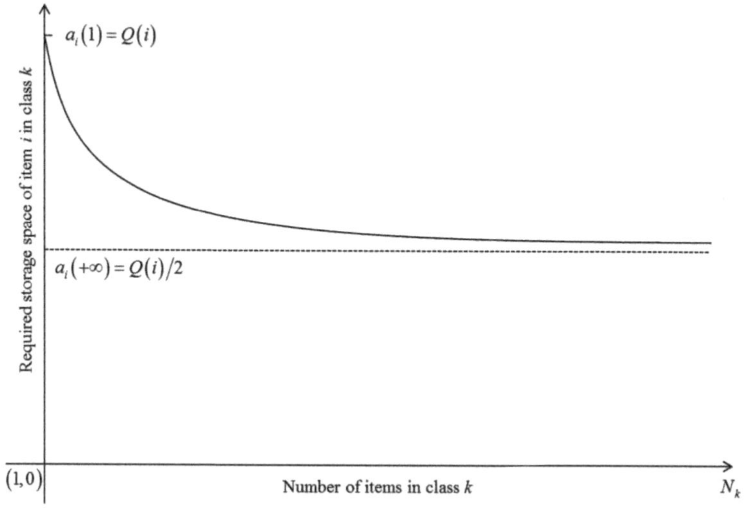

Fig. 2.4 Required storage space of item i as a function of N_k

To provide a better understanding of the above mentioned relationship, we develop an analytical model for a special case by assuming that N_k identical items exist in class k. If the replenishments for the N_k items are fully coordinated, that is, item 1 arrives at time T, item 2 at time $T + T/N_k$, ..., item i at time $T + (i - 1)T/N_k$, and so on, then the inventory patterns of the items are according to those shown in Fig. 2.5. Therefore, the total required storage space of this class, k, can be obtained as the total inventory level of the N_k items when an order arrives. In other words, $\sum_{i=1}^{N_k} i\,Q(i)/N_k = (1 + N_k)Q(i)/2$. Consequently, the required storage space for item i can be obtained as follows:

$$a_i(N_k) = ((1 + N_k)Q(i)/2) \,/\, N_k = 0.5\left(1 + N_k^{-1}\right) Q(i). \tag{2.11}$$

In this special case, the space-sharing factor is $\varepsilon = 1$, which yields the best space-sharing because of the item symmetry and synchronization of the item ordering. However, in practice, the value of ε is significantly smaller than 1 because of the heterogeneity of the items in demand volumes, reorder points, order quantities, delivery lead times, holding costs, and others. Therefore, we use simulation to determine the

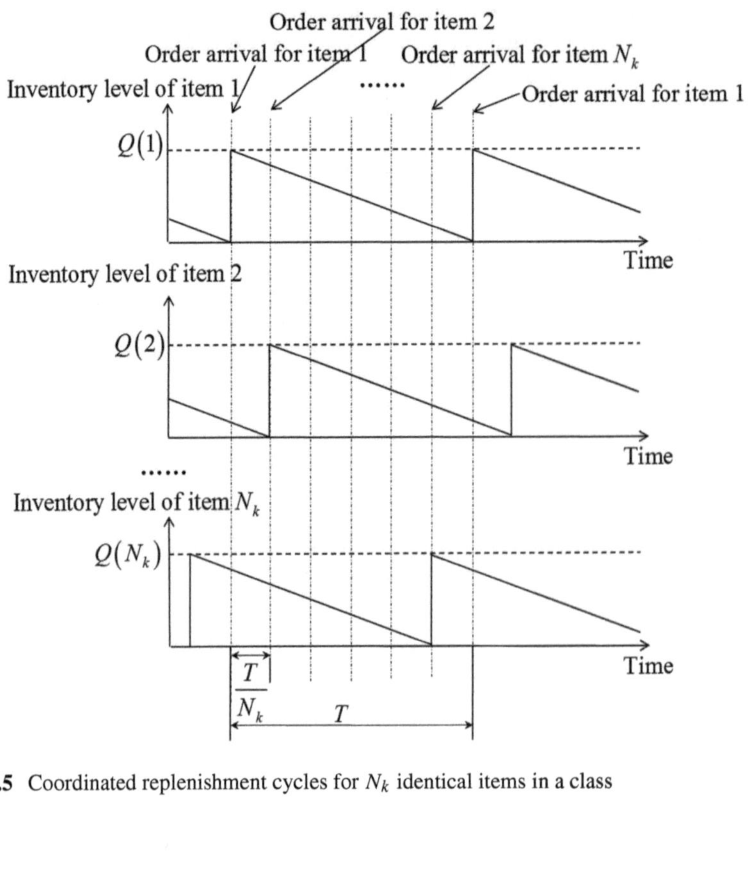

Fig. 2.5 Coordinated replenishment cycles for N_k identical items in a class

Table 2.2 Algorithm used to simulate ε

Steps	Job description
Step 1	Obtain the initial inventory level for the N_k items randomly from their possible values
Step 2	Obtain the inventory level as a function of time for all the N_k items
Step 3	Obtain the total inventory level of this class as a function of time based on the results of *Step 2*
Step 4	Obtain the maximum value of the function obtained from *Step 3* as the required storage space of this class
Step 5	Obtain the simulated required storage space for each item according to their weighted contribution to the total required storage space of the class
Step 6	Repeat *Steps 1–5* for m times, and obtain an average value as the required storage space for the items to estimate ε
Step 7	Repeat *Steps 1–6* for every possible N_k
Step 8	Estimate ε for Eq. (2.10) using least squares method based on the simulated values

average value of ε. The algorithm steps shown in Table 2.2 are used in the simulation, and the details are given in Appendix A.

The method of simulation does not depend on the demand curves or replenishment policies. In the details presented in Appendix A, the continuous review (r,S) policy, classic EOQ policy, and the different kinds of ABC demand curves are verified. The estimated average value of ε appears to be mostly in the range of 0.15–0.25 and is quite insensitive to the system parameters.

Consequently, using Eq. (2.10), we can obtain the required storage space for class k with a finite number of items

$$j_k - j_{k-1} = \int_{i_{k-1}}^{i_k} a_i(N_k)di = \int_{i_{k-1}}^{i_k} \left(1 + N_k^{-\varepsilon}\right)\sqrt{0.5KD(i)}di, \tag{2.12}$$

where j_k is the storage location farthest from the depot in class k, which also corresponds to the total required storage space of items 1 to i_k, and $j_0 = 0$.

To make our result comparable with those of the conventional model, we also rescale j by $j * L$, where $L = \sqrt{2Ks}/(s+1)$ (see Eq. (2.7)) according to Hausman et al. (1976) and Rosenblatt and Eynan (1989). Thereafter, using Eqs. (2.2) and (2.12), we obtain the total required storage space for the first k classes as follows:

$$j_k = i_k^{(s+1)/2} + \sum_{l=1}^{k} N_l^{-\varepsilon}\left(i_l^{(s+1)/2} - i_{l-1}^{(s+1)/2}\right). \tag{2.13}$$

Furthermore, the relationship between R_k and j_k in a SIT continuous storage system is $R_k = \sqrt{j_k}$ (Hausman et al. 1976; Rosenblatt and Eynan 1989). Therefore, the boundary of the kth class, R_k, can now be rewritten as follows:

$$R_k = \sqrt{i_k^{(s+1)/2} + \sum_{l=1}^{k} N_l^{-\varepsilon} \left(i_l^{(s+1)/2} - i_{l-1}^{(s+1)/2} \right)}, \tag{2.14}$$

where N_l is the number of items in class l, which is obtained with $N_l = N(i_l - i_{l-1})$, and N is the total number of items in the system.

2.3.2 Basic Travel-Time Model with a Finite Number of Items

Considering the relationship between R_k and i_k shown in Eq. (2.14), we obtain the new model (hereafter called "Model BM" to differentiate the basic model from the extensions in the following sections) that enables us to determine the optimal class boundaries of the continuous AS/RS.

Model BM:

$$\text{Min} \quad T_n = \sum_{k=1}^{n} \frac{2\left(R_k^3 - R_{k-1}^3\right)}{3\left(R_k^2 - R_{k-1}^2\right)} \left(i_k^s - i_{k-1}^s \right), \tag{2.15}$$

$$\text{s.t.} \quad N(i_k - i_{k-1}) \geq 1, \tag{2.16}$$

$$R_k = \sqrt{i_k^{(s+1)/2} + \sum_{l=1}^{k} N_l^{-\varepsilon} \left(i_l^{(s+1)/2} - i_{l-1}^{(s+1)/2} \right)},$$

$$\text{d.v.} \quad R_k,\ k = 1, 2, \ldots, n; \ \text{or}\ i_k, \quad k = 1, 2, \ldots, n-1,$$

where $R_0 = i_0 = 0$ and $i_n = 1$ are known.

Model BM differs from Model CM in three respects. (i) Eq. (2.14) indicates that the class boundary, R_k, is not only related to the last item and the items' demand but also to the number of items in class k and all its preceding classes $1, \ldots, k-1$. (ii) Constraints (2.16) are required to ensure that at least one item is stored in each class because the items and the total required storage space are finite. (iii) The total required storage space of the system, R_n, is an unknown value in Model BM because R_n relates to R_k, $k = 1, 2, \ldots, n-1$, but $R_n = 1$ is known in Model CM. To determine an efficient manner of solving this problem, we introduce a solution methodology based on dynamic programming.

2.3.3 Solution Methodology

Unfortunately, the methodology for solving Model CM used by Rosenblatt and Eynan (1989) cannot be applied because an iterative relation similar to Eq. (2.9) does not hold in this case. In addition, the objective function (2.15) is nonlinear, and we do not know if it is a convex function of R_k (or i_k), $k = 1, 2, \ldots, n$. In addition, constraints (2.16) are nonlinear functions of R_k, $k = 1, 2, \ldots, n$, in considering Eq. (2.14). For a small number of classes, a grid search can be applied to identify approximate solutions. We use a different solution method because we are interested in a fast algorithm that determines the optimal solution for a larger number of classes. Rewriting expressions (2.14) and (2.15) is possible to enable the use of a dynamic programming solution approach. We denote $Y_k = \sum_{l=1}^{k} N_l$ as the cumulative number of items of the first k classes, where N_l is the number of items in the lth class. The relationship between i_k and Y_k is as follows:

$$i_k = Y_k/N. \tag{2.17}$$

By substituting Eq. (2.17) into Eqs. (2.14) and (2.15), we can rewrite Model BM in a Solution Model (SM) as follows:

Model SM:

$$Min\ \ T_n = \sum_{k=1}^{n} \frac{2\left(R_k^3 - R_{k-1}^3\right)}{3\left(R_k^2 - R_{k-1}^2\right)} \left(\left(\frac{Y_k}{N}\right)^s - \left(\frac{Y_{k-1}}{N}\right)^s\right), \tag{2.18}$$

$$s.t.\ \ Y_k = \sum_{l=1}^{k} N_l,$$

$$R_k = \sqrt{\left(\frac{Y_k}{N}\right)^{(s+1)/2} + \sum_{l=1}^{k} N_l^{-\varepsilon} \left(\left(\frac{Y_l}{N}\right)^{(s+1)/2} - \left(\frac{Y_{l-1}}{N}\right)^{(s+1)/2}\right)}, \tag{2.19}$$

$$d.v.\ \ N_k > 0,\ k = 1, 2, \ldots, n.$$

The details of the dynamic programming solution methodology of this model are described in Appendix B. The complexity of the algorithm appears to be $O(N^3)$.

2.4 Model Extensions

This section extends the basic model in Sect. 2.3 to consider an NSIT storage rack, a different ABC demand curve, a discrete rack, and a stochastic item demand. In particular, Sect. 2.4.1 considers the NSIT rack model, Sect. 2.4.2 examines the model based on Bender's ABC demand curve, and Sect. 2.4.3 extends the model to a discrete storage rack with stochastic item demand.

2.4.1 NSIT Storage Racks

In practice, storage racks are usually NSIT. Following Eynan and Rosenblatt (1994), we discuss an NSIT case with a rack face with a fixed height, $\sqrt{b}$, in the vertical direction, where b is a shape factor with $0 < b \le 1$.

When class-based storage is applied, three types of class regions exist for the NSIT storage rack: square regions, rectangular regions, and a transient region as shown in Fig. 2.6. Square regions like class 1 or L-shaped regions with square outer boundaries (surrounding a square class 1). The boundary of such a region is $R_k \le \sqrt{b}$, and the total area containing the previous k classes is $R_k \times R_k$. Rectangular regions like class 3 with $R_{k-1} \ge \sqrt{b}$. The total area containing the previous k classes is $R_k \times \sqrt{b}$. A transient region is like class 2. It is a region between the square area and the rectangular area. For this class, $R_{k-1} \le \sqrt{b}$ and $R_k \ge \sqrt{b}$. The total area containing the previous k classes is $R_k \times \sqrt{b}$. We note that this region may not exist if the boundary of the $(k-1)$th region is exactly at $R_{k-1} = \sqrt{b}$. According to Eynan and Rosenblatt (1994) and with class $\hat{k}$ as the transient region, the following are true:

$$R_k = \begin{cases} \sqrt{j_k}, & \text{if } 1 \le k < \hat{k}, \\ j_k/\sqrt{b}, & \text{if } k \ge \hat{k}. \end{cases}$$

$$t_k = \begin{cases} 2\left(R_k^3 - R_{k-1}^3\right) / \left(3R_k^2 - 3R_{k-1}^2\right), & \text{if } 1 \le k < \hat{k}, \\ \left(b^{3/2} + 3\sqrt{b}R_k^2 - 4R_{k-1}^3\right) / \left(6\sqrt{b}R_k - 6R_{k-1}^2\right), & \text{if } k = \hat{k}, \\ \left(R_k + R_{k-1}\right)/2, & \text{if } k > \hat{k}. \end{cases}$$

Therefore, the travel-time model with a finite number of items in an NSIT system can be obtained as follows:

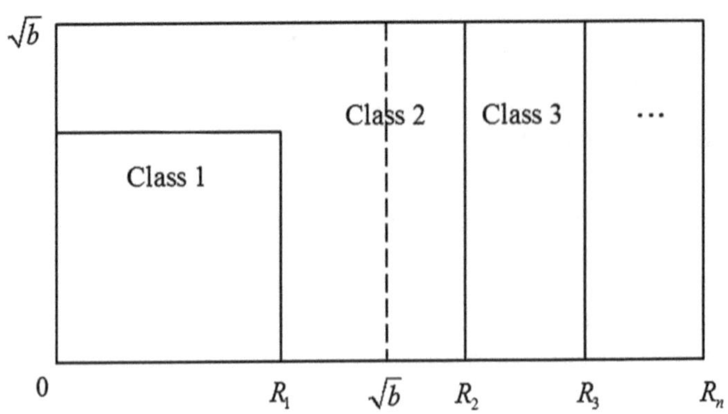

Fig. 2.6 Different types of classes in a NSIT warehouse

Model NSIT:

$$Min \quad T_n = \sum_{k=1}^{\hat{k}-1} \frac{2\left(R_k^3 - R_{k-1}^3\right)}{3\left(R_k^2 - R_{k-1}^2\right)}\left(i_k^s - i_{k-1}^s\right) + \sum_{k=\hat{k}+1}^{n} \frac{(R_k + R_{k-1})}{2}\left(i_k^s - i_{k-1}^s\right)$$

$$+ \frac{\left(b^{3/2} + 3\sqrt{b}R_k^2 - 4R_{k-1}^3\right)}{6\left(\sqrt{b}R_k - R_{k-1}^2\right)}\left(i_k^s - i_{k-1}^s\right), \tag{2.20}$$

$$s.t. \quad Y_k = \sum_{l=1}^{k} N_l,\ i_k = Y_k/N,$$

$$j_k = i_k^{(s+1)/2} + \sum_{l=1}^{k} N_l^{-\varepsilon}\left(i_l^{(s+1)/2} - i_{l-1}^{(s+1)/2}\right),$$

$$R_k = \sqrt{j_k}, \text{ for } 1 \le k < \hat{k},$$

$$R_k = j_k/\sqrt{b}, \text{ for } k \ge \hat{k},$$

$$d.v. \quad N_k > 0,\ k = 1, 2, \ldots, n.$$

The optimal T_n and $N_k > 0, k = 1, 2, \ldots, n$ can be found through the solution methodology presented in Sect. 2.3.3.

2.4.2 Bender's ABC Demand Curve

Although the ABC demand curve in Eq. (2.1) is widely used in the literature, Bender (1981) empirically showed that the following equation well represents the ABC demand curve in reality (Pohl et al. 2011).

$$G(i) = (1 + B)i/(B + i). \tag{2.21}$$

where B is the shape factor of the ABC demand curve. The difference of these two demand curves can be clearly observed in Fig. 2.7 by choosing the 20 %/70 % curve as an example. To examine the effect of these different ABC demand curves on the optimal number of classes, we revise Model BM described in Sect. 2.3 as follows:

With regard to the difference between the two curves, the corresponding item demand, $D(i)$, and the cumulative required storage space for the first k classes, j_k, can be realized as the following:

$$D(i) = B(1 + B)/(B + i)^2, \tag{2.22}$$

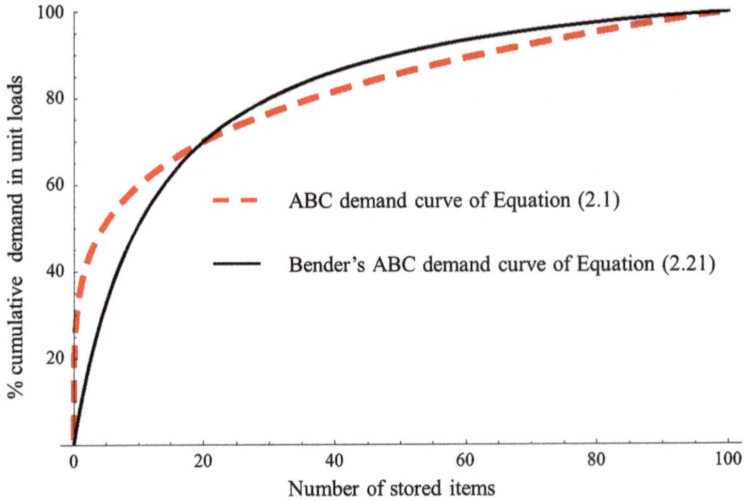

Fig. 2.7 Two different 20 %/70 % ABC demand curves

$$j_k = \left(\ln \frac{B + i_k}{B} + \sum_{l=1}^{k} \ln \frac{B + i_l}{B + i_{l-1}} \right) \Big/ \ln \frac{B+1}{B}. \tag{2.23}$$

Therefore, by replacing i_k with Y_k/N and with respect to $R_k = \sqrt{j_k}$, we obtain the model for Bender's ABC demand curve Model SMB ("SMB" represents "Solution Methodology with Bender's ABC curve") as follows:

Model SMB:

$$\textit{Min} \quad T_n = \sum_{k=1}^{n} \frac{2\left(R_k^3 - R_{k-1}^3\right)}{3\left(R_k^2 - R_{k-1}^2\right)} \left(\frac{(1+B)Y_k}{NB + Y_k} - \frac{(1+B)Y_{k-1}}{NB + Y_{k-1}} \right), \tag{2.24}$$

$$s.t. \quad Y_k = \sum_{l=1}^{k} N_l,$$

$$R_k = \sqrt{\left(\ln \frac{B + i_k}{B} + \sum_{l=1}^{k} \ln \frac{B + i_l}{B + i_{l-1}} \right) \Big/ \ln \frac{B+1}{B}}, \tag{2.25}$$

$$d.v. \quad N_k > 0, \; k = 1, 2, \ldots, n.$$

Through Model NSIT, the optimal T_n and $N_k > 0, k = 1, 2, \ldots, n$ can be obtained similarly through the same solution methodology shown in Sect. 2.3.3.

2.4.3 Discrete Racks and Stochastic Demand

In this section, our basic model (Model BM) for continuous racks and deterministic demand in Sect. 2.3 is extended to a more realistic model with discrete storage locations and stochastic item demand. Different from the problem described in Sect. 2.2, the demand of each item follows a stochastic distribution over a unit-time period. The expected demand is determined according to the ABC demand curve. Item inventories are replenished according to a continuous review (r,S) policy with a replenishment lead time. Here, r is the reorder point, and S is the order-up-to level.

Shortages may occur and lead to backorders because item demand is stochastic and a delivery lead time exists for replenishing orders. Therefore, a maximum stockout probability is set. The stockout probability of item i in the warehouse must be less than α_i, with $(1-\alpha_i)$ defined as the service level (or fill rate). As a result, a safety stock ss_i for item i is needed to achieve this service level.

Correspondingly, in addition to the notations given in Sect. 2.2, the notations shown in Table 2.3 are defined for this model.

To obtain the travel-time model for an n-class-based storage system shown as Eq. (2.6), we need to get the average travel time for each storage class and the weighted expected turnover frequency of each class.

First, the cumulative fraction of the total expected demand of the i items can be expressed as follows according to Hausman et al. (1976):

$$G(i) = (i/N)^s = \sum_{x=1}^{i} \lambda(x) \Big/ \sum_{x=1}^{N} \lambda(x), \quad 0 < s \leq 1. \tag{2.26}$$

Second, we define τ_j as the one-way travel time from location j to the depot. For a discrete SIT system, τ_j can be expressed as follows:

$$\tau_j = \left\lceil \sqrt{j} \right\rceil, \tag{2.27}$$

Table 2.3 Notations used in Sect. 2.4.3

Notations	Definitions
r_i	Reorder point of item i
S_i	Order-up-to level of item i
l_i	Delivery lead time for the orders of item i, which is constant for each item
$\lambda(i)$	Expected demand (i.e., expected number of retrievals) of item i in a unit-time period with $\lambda(i) \geq \lambda(i+1)$ for all i. The expected demand of the item over the delivery lead time is $l_i\lambda(i)$
$f_i(\cdot)$	Demand probability density function of item i during the delivery lead time l_i
$F_i(\cdot)$	Cumulative demand distribution function of item i during the delivery lead time l_i

where $\lceil \sqrt{j} \rceil$ represents the smallest integer number not less than $\sqrt{j}$, and the travel speed is supposed to be one location per second. As a result, the average one-way travel time for storing (or retrieving) a unit load of class k can be obtained as follows:

$$t_k = \frac{1}{j_k - j_{k-1}} \sum_{j=j_{k-1}+1}^{j_k} \tau_j. \tag{2.28}$$

Therefore, similar to Eq. (2.6), the average travel time for the system can be obtained as follows:

$$T_n = \sum_{k=1}^{n} \frac{\sum_{j=j_{k-1}+1}^{j_k} \tau_j}{j_k - j_{k-1}} \left(\left(\frac{Y_k}{N} \right)^s - \left(\frac{Y_{k-1}}{N} \right)^s \right). \tag{2.29}$$

With the description given so far, the problem now is to determine the relationship between j_k and Y_k. First, the function of the required storage space of item i, $a_i(N_k)$ for determining $j_k - j_{k-1}$ can be obtained through simulations similar to Model BM as follows:

$$a_i(N_k) = 0.5 \left(1 + N_k^{-\varepsilon}\right)(S_i - r_i) + ss_i. \tag{2.30}$$

Thereafter, the required storage space for class k can be obtained as follows:

$$\begin{aligned} j_k - j_{k-1} &= \left\lceil \sum_{i=Y_{k-1}+1}^{Y_k} a_i(N_k) \right\rceil \\ &= \left\lceil \sum_{i=Y_{k-1}+1}^{Y_k} \left(0.5\left(1 + N_k^{-\varepsilon}\right)(S_i - r_i) + ss_i\right) \right\rceil, \end{aligned} \tag{2.31}$$

where $j_0 = 0$.

Consequently, the total required storage space of the first k classes is equal to the following:

$$j_k = \left\lceil \sum_{l=1}^{k} \sum_{i=Y_{l-1}+1}^{Y_l} \left(0.5\left(1 + N_l^{-\varepsilon}\right)(S_i - r_i) + ss_i\right) \right\rceil, \tag{2.32}$$

where $N_l = Y_l - Y_{l-1}$ and $Y_0 = 0$.

Considering the previous analysis, the relationship between j_k and Y_k expressed in Eq. (2.32), and the objective function given in Eq. (2.29), we obtain the following Model DSM (i.e., discrete-stochastic model for discrete racks and stochastic demand) to determine the optimal class allocations of the AS/RS in the discrete space scenario.

Model DSM:

$$\begin{aligned} &Min \quad T_n = \sum_{k=1}^{n} \frac{\sum_{j=j_{k-1}+1}^{j_k} \tau_j}{j_k - j_{k-1}} \left(\left(\frac{Y_k}{N}\right)^s - \left(\frac{Y_{k-1}}{N}\right)^s \right), \qquad (2.33) \\ &s.t. \quad \tau_j = \left\lceil \sqrt{j} \right\rceil, \quad Y_k > Y_{k-1} \text{ and Eq. (2.32)}, \\ &d.v. \quad Y_k > 0, \ k = 1, 2, \ldots, n, \end{aligned}$$

where $j_0 = Y_0 = 0$ is known.

Similarly, this model can be solved by the methodology provided in Sect. 2.3.3.

2.5 Numerical Illustrations

This section provides detailed numerical illustrations to present in detail the results of the models deduced in this chapter and to show managerial insights to help warehouse managers make useful decisions. In Sect. 2.5.1, the results of Model CM from Sect. 2.2 and those of our new Model BM from Sect. 2.3 are compared under different ABC demand curves. Section 2.5.2 presents the results of the extended models with respect to NSIT storage racks, Bender's ABC demand curve, and discrete rack and stochastic item demand.

2.5.1 Base Examples: Results for Basic Model

In our base example, the total number of items in the system is $N = 100$, and the space-sharing factor is $\varepsilon = 0.22$ (the average value, obtained through simulation, see Appendix A). This section shows the comparison between Model BM and Model CM for different numbers of classes. We solve both models for 1 class to 100 classes. The results for the optimal travel time T_n as a function of the number of classes n are shown in Fig. 2.8 under four different ABC demand curves, with $s = 1(20\,\%/20\,\%)$, $0.431(20\,\%/50\,\%)$, $0.222(20\,\%/70\,\%)$, and $0.065(20\,\%/90\,\%)$. The corresponding required storage space of the system is shown in Fig. 2.9.

The results shown in Figs. 2.8 and 2.9 imply the following:

(i) The optimal number of classes n^* is small, and $n = 3$ provides near-optimal solutions in all cases tested. In Fig. 2.8, $n^* \leq 5$ for all our examples, which cover all practical values of the ABC demand curves. A small number of classes is very close to the warehousing practice, in which only three storage classes are often used (Roodbergen and Vis 2009).

(ii) Travel time is insensitive to the number of classes in a wide range beyond the optimum. In Fig. 2.8, the differences in travel time between $n = 3$ and $n = 8$ are basically negligible. However, the range of the number of classes yielding the

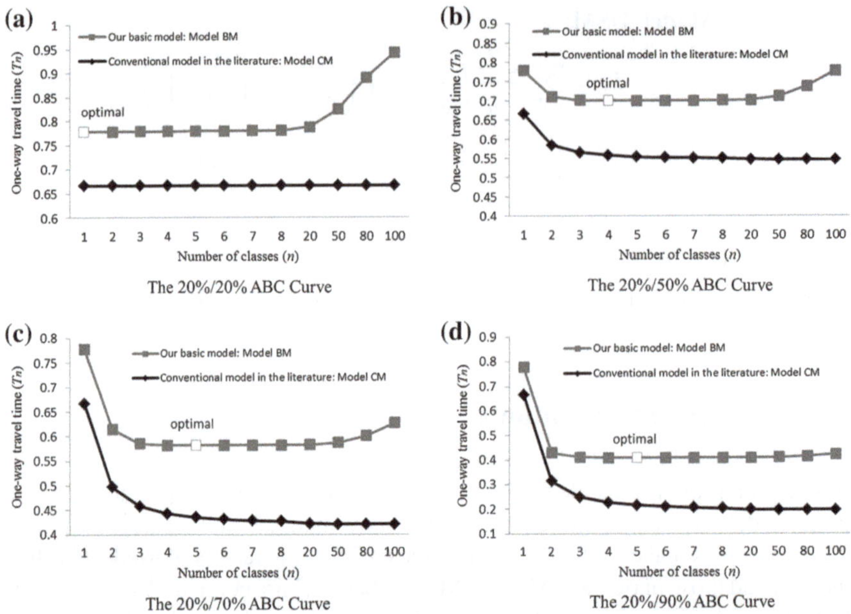

Fig. 2.8 Travel times in our basic model compared to those in the literature. **a** The 20 %/20 % ABC curve. **b** The 20 %/50 % ABC curve. **c** The 20 %/70 % ABC curve. **d** The 20 %/90 % ABC curve.

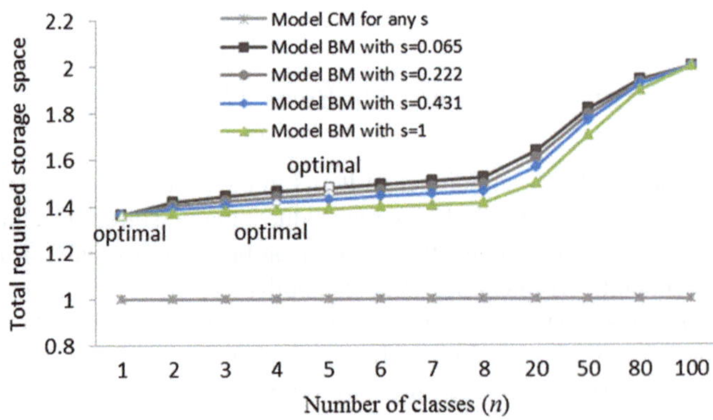

Fig. 2.9 Total required storage space as a function of the number of classes

near-optimal travel time depends on the ABC demand curves. For instance, if we define that the acceptable gap is 1 % for the 20 %/20 % curve, then the range of 1 class to 17 classes is acceptable, whereas for the 20 %/90 % curve, the acceptable range is from 3 classes to 75 classes. If the acceptable gap is 5 %, then the range of 1 class to 44 classes is acceptable for the 20 %/20 % curve and that of 3 classes to 100 classes is all acceptable for the 20 %/90 % curve. That is, a full turnover-based policy is acceptable for this curve. Therefore, managers should not hesitate to select any reasonable small number of classes because the result is essentially near optimal. This result is also true even when the required storage space is considered because the required storage space does not increase significantly from $n = 3$ to 8 as shown in Fig. 2.9.

(iii) The relative gaps between the travel times of our Model BM and those of Model CM increase with n in all examples. Even at $n = 1$, the gaps are still quite large (almost 15 % for all examples in Fig. 2.8). Therefore, warehouse managers should not only simply adopt the average inventory level as the required storage space for the warehousing system. The underestimation of required storage space may lead to managers' incorrect decisions, such as response time estimation and cost budgeting.

(iv) Compared with the required storage space adopted in conventional research, the required storage space of Model BM and that of Model CM have a significant gap that is usually more than 30 % and can be as large as 100 % when dedicated storage is adopted. Consequently, warehouse managers should consider the difference among the required storage spaces when the storage policy changes, and this result provides warehouse managers a direction about warehouse capacity design.

2.5.2 *Results for Extended Models*

To verify the robustness of the results obtained in Sect. 2.5.1, this section presents the numerical results for the extended models given in Sect. 2.4. Noting that the values of parameters are different section between Model DSM and the others, we first show the results for the NSIT storage rack and Bender's ABC curve in Sect. 2.5.2.1 and then provide the results of Model DSM in Sect. 2.5.2.2.

2.5.2.1 Results for the NSIT Storage Racks and Bender's ABC Demand Curve

This section presents the results for the NSIT storage racks and Bender's ABC demand curve. The results shown in Fig. 2.10 are based on a 20 %/70 % curve, which is a common case in practice. The results based on other curves with different skewness are omitted here because they have the same trend.

The results show that the main findings of Sect. 2.5.1 still hold for different storage racks and different kinds of ABC demand curves: a small number of classes yield the minimum travel time of the system, and any number of classes around the optimal one is a near-optimal solution.

In particular, an NSIT storage rack leads to longer travel times than a SIT rack because the optimal rack configuration is not NSIT but SIT (as shown in Fig. 2.10a). In Fig. 2.10b, the travel time of the conventional ABC curve (Eq. (2.1)) is shorter than

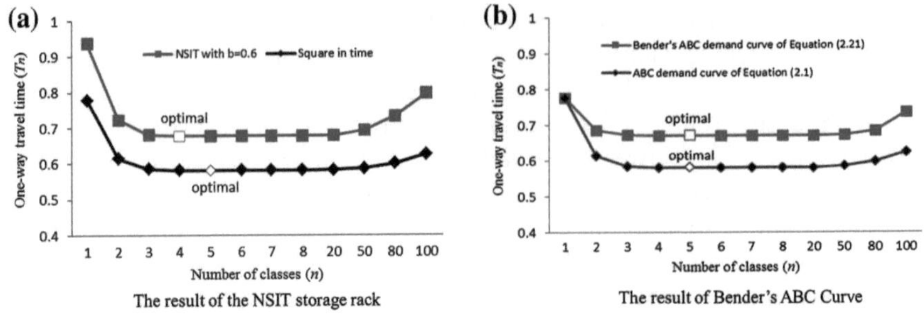

Fig. 2.10 Travel time for Model NSIT and Model SMB. **a** The result of the NSIT storage rack. **b** The result of Bender's ABC curve

that of Bender's ABC curve (Eq. (2.21)) because the conventional ABC demand curve is more skewed (as shown in Fig. 2.7).

2.5.2.2 Results for Discrete Racks and Stochastic Demand

The basic parameters in this section are as follows: $N = 100$ items are stored in the warehousing system; the total annual demand of all the items is $\sum_{i=1}^{N} \lambda(i) = 10,000$; and $K = 2$ and $l_i = 1/50$ year for $i = 1, 2, \ldots, N$. A continuous review order-up-to level replenishment policy is adopted for item i, $i = 1, 2, \ldots, N$ with the required service level $1-\alpha_i = 95\,\%$ for a 20 %/70 % demand curve. The demand of each item follows a normal distribution with $\sigma_i/\mu_i = 0.2$ and $\mu_i = \lambda(i)$ for all $i = 1, 2, \ldots, N$. The reorder point r_i is chosen according to the service level as $r_i = F_i^{-1}(1-\alpha_i)$. Thereafter, the safety stock level can be obtained as $ss_i = r_i - l_i\lambda(i)$. Finally, according to Kapalka et al. (1999), the order-up-to level is $S_i = r_i + \sqrt{2K\lambda(i)}$.

The results of Model DSM for discrete racks and stochastic demand are given in Figs. 2.11, 2.12 and 2.13. The comparison of the results with those of Model BM can be found in Fig. 2.11, in which the results of Model DSM are normalized through the replacement of R_j with $R_j \Big/ \sqrt{\sum_{i=1}^{N} Q(i)/2}$, comparable with the normalized results of Model BM. We only present the result of $s = 0.222$ for an example in Fig. 2.11 because the results of different ABC demand curves are similar. Figures 2.12 and 2.13 show the sensitivity results of the optimal number of classes and the corresponding required storage space with varying parameters.

The results indicate that our major findings in Sect. 2.5.1 hold in the case of discrete rack and stochastic demand; a small number of classes lead to a minimum travel time (as shown in Figs. 2.11a and 2.12) through varying number of items, service level, and demand variability. The minimum travel time is insensitive to the optimal number of classes around the optimum.

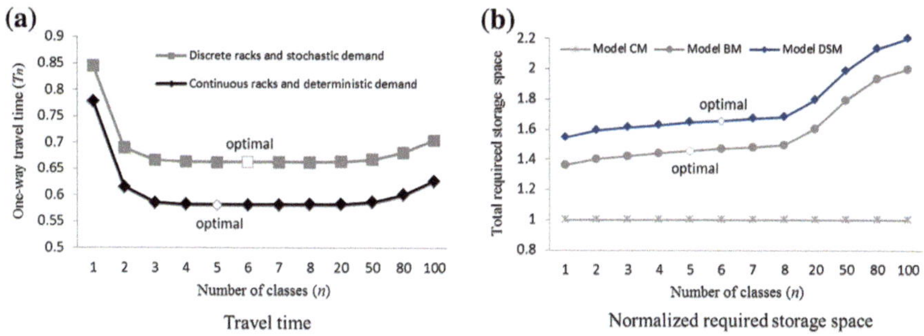

Fig. 2.11 Comparison of Model BM and Model DSM, $s = 0.222$. **a** Travel time. **b** Normalized required storage space

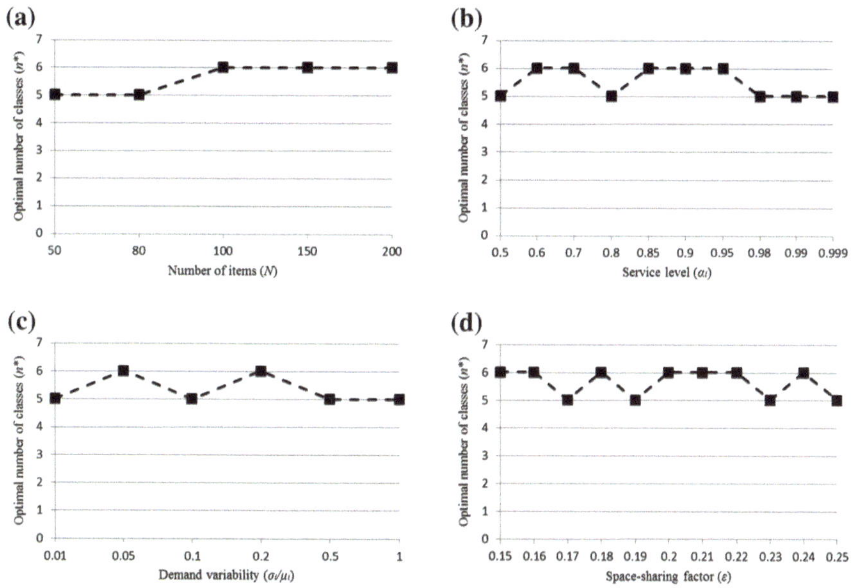

Fig. 2.12 Optimal number of classes, depending on N, α_i, σ_i/μ_i and ε. **a** Number of items. **b** Service level. **c** Demand variability. **d** Space-sharing factor

In particular, as shown in Fig. 2.10a, the travel time of Model DSM is longer than that of Model BM. The difference mainly comes from the extra space needed for the safety stock of the items in Model DSM. Figure 2.10b shows that more required storage spaces for Model DSM are needed, that is, at least 60 % larger than those of Model CM for eight storage classes and even 100 % for a large number of classes.

Figure 2.13 illustrates that the required storage space increases with the required service level, demand variability, and number of items in the system but decreases with the space-sharing factor. The required storage space convexly increases because

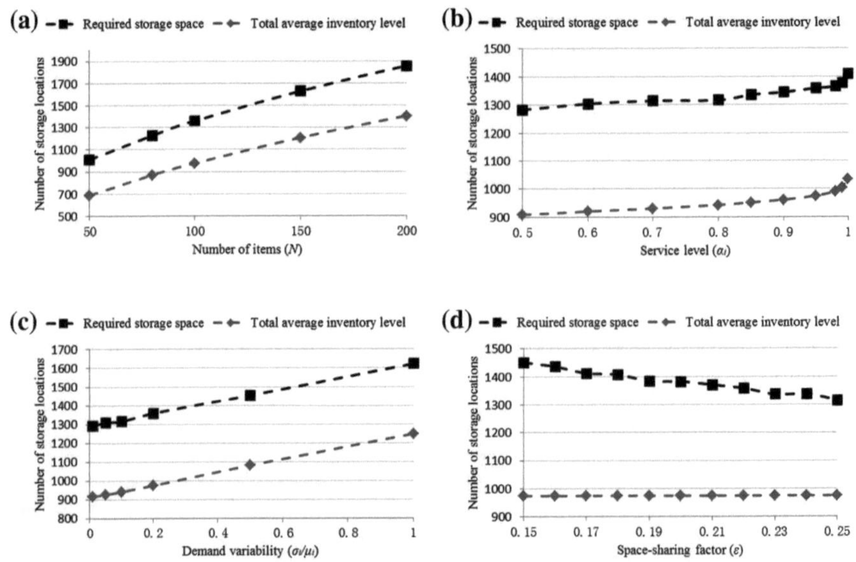

Fig. 2.13 Total required storage space at the optimal number of classes, depending on $N, \alpha_i, \sigma_i/\mu_i$ and ε. **a** Number of items. **b** Service level. **c** Demand variability. **d** Space-sharing factor

the safety stock increases with the service level convexly. The required space is largely linear in the demand variability σ_i/μ_i.

2.6 Chapter Summary

This chapter extends the research on class-based storage by considering a finite number of items to be stored in an AS/RS. Our results reveal that the optimal number of classes is relatively small, and three classes give the near-shortest travel time for the ABC demand curves between 20 %/50 % and 20 %/90 %. This finding contradicts the common idea in the literature that more storage classes are better in view of travel time.

The results also show a flat range of the number of classes yielding the near-optimal solution, which is dependent on the skewness of the demand curves. For example, for the 20 %/20 % curve, 17 classes lead to a 1 % increase in travel time compared with the minimum value, and 45 classes lead to more than 5 %. For the 20 %/90 % curve, a 1 % gap enables managers to adopt up to 75 classes, and the gap is less than 3.5 % for the 100-class system. Although the flat range is influenced by the demand curves, a common range provides a near-optimal solution for all the cases. This finding is important for warehouse managers because it suggests that any reasonable number of classes such as between 3 and 8 is near optimal. As a result, warehouse managers can freely change the number of classes when necessary (e.g., for space use purposes) because travel time is not sensitive to the number of classes.

Furthermore, by analyzing the extended models, we show that the results are robust for different storage rack shapes, different kinds of ABC demand curves, discrete storage locations, and stochastic item demand. These findings provide important managerial directions about item classification when class-based storage is adopted in a warehousing system.

Moreover, we reveal that the space needed for a warehouse with an optimal number of storage classes should be at least 30–50 % of the average inventory level. This result is in accordance with the practical knowledge that every warehouse needs a slack space to warrant a smooth operation. An important conclusion for the research is that the space-sharing effect cannot be ignored in the class-based storage system. Models assuming perfect space-sharing (i.e., by implicitly assuming an infinite number of items per class) underestimate the space requirements, the travel time needed, and investments in racks and equipment. This finding provides a clearer guidance for capacity design, instead of simply adopting the average inventory level, in a class-based storage system.

Therefore, further research on storage policies based on a finite number of items is called for because many studies use class-based storage and implicitly assuming an infinite number of items. The results of these studies should be revisited to address the consequence of assuming a finite number of items to be stored in the system. The results for a finite number of items may substantially differ from those for an infinite number of items. First, travel time results under the assumption of an infinite number of items per class are usually overly optimistic because a division in storage classes requires more storage space than that accounted for. Second, an increasing number of storage classes increases the response times rather than reducing them. Similar results will hold for parallel-aisle or fishbone-layout warehouses (Gue and Meller 2009). Our contribution to identify the tradeoff between travel time reduction by item ranking and increase through increased storage space leading to an optimal product and storage number of classes may also be applied to other areas where products are classified by some criterion.

References

Adams ND (1996) Warehouse and distribution automation handbook. McGraw-Hill Companies, New York

Bender PS (1981) Mathematical modeling of the 20/80 rule: theory and practice. J Bus Logist 2(2):139–157

Bozer YA, Schorn EC, Sharp GP (1990) Geometric approaches to solve the chebyshev traveling salesman problem. IIE Trans 22(3):238–254

De Koster R, Le-Duc T, Roodbergen KJ (2007) Design and control of warehouse order picking: a literature review. Eur J Op Res 182(2):481–501

Eynan A, Rosenblatt MJ (1994) Establishing zones in single-command class-based rectangular as/rs. IIE Trans 26(1):38–46

Gu J, Goetschalckx M, McGinnis LF (2007) Research on warehouse operation: a comprehensive review. Eur J Op Res 177(1):1–21
Gue KR, Meller RD (2009) Aisle configurations for unit-load warehouses. IIE Trans 41(3):171–182
Hausman WH, Schwarz LB, Graves SC (1976) Optimal storage assignment in automatic warehousing systems. Manag Sci 22(6):629–638
Heragu SS (2006) Facilities Design. iUniverse
Johnson ME, Brandeau ML (1996) Stochastic modeling for automated material handling system design and control. Transp Sci 30(4):330–350
Kapalka BA, Katircioglu K, Puterman ML (1999) Retail inventory control with lost sales, service constraints, and fractional lead times. Prod Op Manag 8(4):393
Kouvelis P, Papanicolaou V (1995) Expected travel time and optimal boundary formulas for a two-class-based automated storage/retrieval system. Int J Prod Res 33(10):2889–2905
Larson TN, March H, Kusiak A (1997) A heuristic approach to warehouse layout with class-based storage. IIE Trans 29(4):337–348
Pohl LM, Meller RD, Gue KR (2011) Turnover-based storage in non-traditional unit-load warehouse designs. IIE Trans 43(10):703–720
Roodbergen KJ, Vis IF (2009) A survey of literature on automated storage and retrieval systems. Eur J Op Res 194(2):343–362
Rosenblatt MJ, Eynan A (1989) Note deriving the optimal boundaries for class-based automatic storage/retrieval systems. Manag Sci 35(12):1519–1524
Tompkins JA, White JA, Bozer YA, Tanchoco JMA (2010) Facilities planning. Wiley, New Jersey
Yu Y, de Koster RB (2009) Optimal zone boundaries for two-class-based compact three-dimensional automated storage and retrieval systems. IIE Trans 41(3):194–208

Chapter 3
Performance of Class-Based Storage in a Unit-Load Warehouse

Abstract Storage policy performance in a warehouse is usually evaluated through average one-way travel distance or time. By considering the realistic required storage space (RSS) measured in number of locations, this chapter examines the performance of random, full turnover-based, and class-based storage policies for a unit-load warehouse operated by a forklift in single-command mode with a given annual demand volume. A generalized travel distance model with consideration of realistic RSS is developed to describe the performance problem of these policies. According to this model, the results show that on the one hand, the performance of random policy does not keep flat but increases with the skewness of the demand curve because of the effect of realistic RSS. On the other hand, not turnover-based but class-based policy is recommended to warehouse managers because with the consideration of the tradeoff between turnover ranking and space sharing among items, class-based policy performs better than full turnover-based policy. In addition, warehouse managers should adopt a wide-shallow warehouse layout for items with nonsignificantly different annual demands and a narrow-deep layout for items with significant differences because the optimal warehouse shape factor (ratio of width to depth) decreases with the skewness of the demand curve.

3.1 Research Background

The required storage space (RSS) of items is an important term in warehousing management, including warehouse design, storage policy selection, and one-way travel distance evaluation. The existing research considers the RSS of an item to be equal to its corresponding average inventory level and further divides it by the total average inventory level of all the items stored in the warehouse for convenience (Adams 1996; Ashayeri et al. 2002; Koster et al. 2007; Graves et al. 1977; Koh et al. 2002; Kouvelis and Papanicolaou 1995; Thonemann and Brandeau 1998; Tompkins et al. 2010; Berg 1996; Yu and Koster 2009; Zaerpour et al. 2013). Some well-known results are inducted on the basis of this assumption such as the following: the average one-way travel time for the S/R machine to retrieve a unit-load in a SIT AS/RS under random policy is 2/3, and it is not affected by the demand structure of the stored items

X. Guo, *Storage Policies and Maintenance Support Strategies in Warehousing Systems*, Springer Theses, DOI 10.1007/978-981-10-1448-2_3

(i.e., skewness of the demand curve) (Hausman et al. 1976). Another famous result is that the full turnover-based policy is the best solution for a warehousing system in achieving the minimal expected one-way travel time. It is also widely used as a benchmark in evaluating the performance of class-based policy (Teunter et al. 2010; Rosenblatt and Eynan 1989).

In addition to the literature on AS/RS, storage policies are also examined and similar results are obtained in aisle-based traditional warehouses (Jarvis and McDowell 1991; Petersen and Schmenner 1999; Petersen and Aase 2004; Petersen et al. 2004; Koster et al. 2012; Roodbergen 2012). For example, Le-Duc and Koster (2005) optimized the classification based on a closed-form travel time estimation for a two-block discrete warehouse. Petersen et al. (2004) examined the improvement of order-picking performance with the implementation of a class-based storage policy compared with a random policy. Rao and Adil (2013) studied the optimal class boundaries for two- to four-class-based low-level order-picking systems. They provided the corresponding optimal class boundaries, number of aisles, and pick list size for the traditional warehouse with a two-block layout and low-level aisles. Petersen and Schmenner (1999) evaluated three variations of class-based storage policy, diagonal storage, within-aisle storage, and across-aisle storage in an order-picking operation.

However, as shown in Chap. 2, the RSS for an item in practice is higher than its average inventory level because of the limited number of items in a warehouse (Yu et al. 2015). This issue raises a realistic challenge to the well-known results in existing literature. The realistic RSS measured in number of locations of an item is influenced by its annual demand, replenishment policy, and space-sharing effect with other items in the same warehouse. As a result, the total RSS of a warehouse is influenced by the item demand structures (different ABC demand curves) and space-sharing effects among items (total number of items in the warehouse). Furthermore, the relationship between cumulative item turnover and cumulative RSS should also be changed. As a result, the difference between RSSs in the literature and those in practice yields different findings compared with the well-known results mentioned above. This study aims to verify the influence of realistic RSS, which depends on warehouse demand structures and item space-sharing effect, on storage policy performance and warehouse design in a traditional warehouse with a given annual demand volume.

A generalized travel distance model with consideration of realistic RSS measured in number of locations is developed to obtain the optimal solution of the three policies, namely, random, turnover-based, and class-based policies. Thereafter, the corresponding average one-way travel distances for the policies can be obtained, and the performance can be evaluated according to the travel distances in different situations under ABC demand curves with varying skewness.

First, based on the travel distance model, the computational results indicate that along with the increasing skewness of the demand curve, the average one-way travel distance does not remain constant but decreases for all the three policies because the realistic RSS decreases under a given total demand. Even for the random policy, a significant gap (almost 40 %) exists between the 20 %/20 % and 20 %/90 % demand curves. Second, for the optimal classification (number of classes and corresponding class boundaries) for class-based storage, less than five classes can yield the global minimum average one-way travel distance for a unit-load warehouse considering realistic RSS. Therefore, the performance of the class-based policy is better than that of the full turnover-based policy, which is considered as a perfect benchmark in the literature. Third, the optimal warehouse shape factor (the ratio of warehouse width to warehouse depth) under a class-based policy with optimal item classifications decreases with the skewness of the demand curve. This finding suggests that warehouse managers should adopt a wide-shallow warehouse layout with a large number of short storage aisles for items with similar demand volumes (i.e., flat ABC demand curves with low skewness) and a narrow-deep layout with a small number of long storage aisles for items with significant differences (i.e., quite skewed ABC demand curves with high skewness).

3.2 Problem Description

This section first describes the system investigated in this study. Then, the research problem is introduced and the related notations are provided. We consider a unit-load warehouse consisting of a unit-capacity forklift that stores or retrieves items and operates in a single-command mode, storage racks arranged to follow a parallel-aisle layout, one front aisle where the forklift moves horizontally, and one depot located in the middle of the front aisle where all the items leave and enter the warehouse. The details are illustrated in Fig. 3.1.

In particular, all storage locations in the warehouse are assumed to be normalized to the same size as the unit loads, and each location only stores one item. Furthermore, because every command should take the width distance of the front aisle, we omit

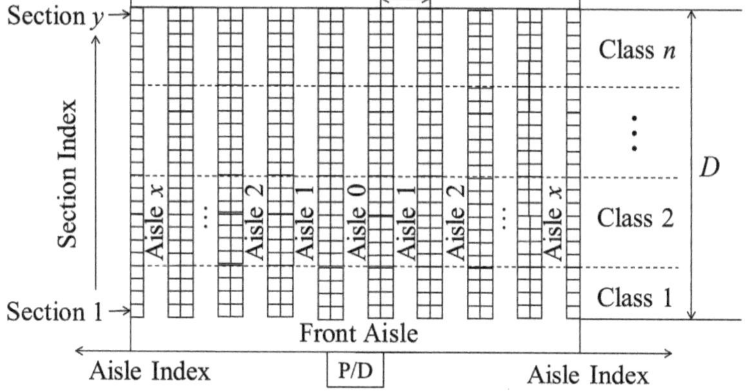

Fig. 3.1 Warehouse system and class-based storage

it from the model (i.e., we assume that the width of the front aisle is 0) without influencing the results. For an easier understanding of the warehouse system and the convenience of presenting the problem, the storage aisle in front of the depot is indexed as Aisle 0, and the index number increases along the right and left sides because the configurations are the same on both sides (as shown in Fig. 3.1). The width of the parallel storage aisles (center-to-center distance between adjacent aisles) is denoted as w. As a result and as shown in Fig. 3.1, the warehouse has x aisles horizontally on both sides of Aisle 0 and is y sections deep. Consequently, with each operation beginning and ending at the depot, the travel distance for each command to visit location (a, b) takes aw horizontally to reach the target aisle and b distance units in depth to reach the target location.

Similar to the method in Chap. 2, N items are stored in the warehouse, and the turnover frequency (demand per unit-time) of each item is constant and known beforehand as described by an ABC demand curve. Moreover, the classic EOQ model is used as replenishment policy for the items without any influence on the results (Hausman et al. 1976).

The aim of this study for the warehouse described above is to evaluate the performance of the random, class-based, and full turnover-based storage policies. Therefore, the problem is to determine the average one-way travel distance for a command under the three policies. A generalized travel distance model is developed to solve this problem. In particular, the model is based on an n-class-based storage system through which the solutions for a random policy and a full turnover-based policy can be obtained when $n = 1$ and $n = N$, respectively. In addition, across-aisle storage (Petersen 2002) is adopted to describe the class boundaries for class-based and full turnover-based policies (as shown in Fig. 3.1).

The notations related to this chapter are introduced and defined in Table 3.1.

With the notations given so far, according to previous studies (Hausman et al. 1976; Rosenblatt and Eynan 1989), the travel distance model for an n-class-based storage system can be written as follows:

$$T_n = \sum_{k=1}^{n} t_k \left(\frac{\Lambda(k)}{\sum_{k=1}^{n} \Lambda(k)} \right), \quad k = 1, 2, \ldots, n, \tag{3.1}$$

where $\Lambda(k) / \sum_{k=1}^{n} \Lambda(k)$ is the weighted turnover frequency of class k in the warehouse. The problem is to determine the relationship between the average one-way travel time in each class and the corresponding total turnover with consideration of realistic RSS. Therefore, the next section discusses the detailed model, and then, the optimal solutions for the random, class-based, and full turnover-based policies can be obtained.

Table 3.1 Notations used in this chapter

Notations	Definitions
$i \in \{1, 2, \ldots, N\}$	Index of the ith item. A smaller index means a larger item turnover
$b \in N^+$	Index of the bth section, numbered from the depot (front aisle)
n	Number of classes that the warehouse wants to apply
k	Index of classes, $k = 1, 2, \ldots, n$
i_k	Item index with the lowest turnover in class k. It also represents the total number of items in the first k classes, $i_0 = 0$
b_k	section index farthest from the depot in class k, the further boundary of class k, $b_0 = 0$
t_k	Average one-way travel distance for storing or retrieving a unit load of class k
R_k	Required storage space of class k, $R_0 = 0$
L_k	Required storage space of the first k classes, $L_0 = 0$
$\Lambda(k)$	Total turnover frequency of class k in number of unit loads per unit-time period of all items stored in class k
T_n	Average one-way travel distance of a unit load for an n-class-based warehouse

3.3 Travel Distance Model Considering Realistic RSS

This section discusses the detailed travel distance model for an n-class-based system in the described unit-load warehouse. As mentioned in the properties proposed above, an ABC demand curve is a plot of ranked cumulative percentage expected demand per unit time. According to the well-known ABC demand curve used in the literature (Hausman et al. 1976), the ABC curve for discrete items can be expressed as follows:

$$G(i) = (i/N)^s = \sum_{j=1}^{i} D(j) \Big/ \sum_{j=1}^{N} D(j), \quad \text{for } 0 < s \leq 1,\ i = 1, 2, \ldots, N, \tag{3.2}$$

where N is the total number of items in the warehouse, $D(i)$ is the annual demand of item i, and s is the shape factor of the ABC demand curve. Without loss of generality and with the annual demand of the warehouse denoted as $A = \sum_{j=1}^{N} D(j)$, the demand of item i can be obtained as follows:

$$D(i) = A\left((i/N)^s - ((i-1)/N)^s\right), \quad i = 1, 2, \ldots, N. \tag{3.3}$$

Consequently, according to the demands of the items, the weighted turnover of class k in the total turnover of all the items in the warehouse, $\Lambda(k) \Big/ \sum_{k=1}^{n} \Lambda(k)$, can be obtained as follows:

$$\frac{\Lambda(k)}{\sum_{k=1}^{n}\Lambda(k)} = \frac{\sum_{j=i_{k-1}+1}^{i_k} D(j)}{\sum_{j=1}^{N} D(j)} = \left(\frac{i_k}{N}\right)^s - \left(\frac{i_{k-1}}{N}\right)^s, \quad i = 1, 2, \ldots, N. \quad (3.4)$$

where $i_0 = 0$. Furthermore, according to the EOQ replenishment policy and annual demand, the order quantity (in unit loads) of item i is as follows:

$$Q(i) = \sqrt{2KD(i)} = \sqrt{2KA\left((i/N)^s - ((i-1)/N)^s\right)}, \quad (3.5)$$

where K is the ratio of reorder cost to holding cost. Similar to the value in Chap. 2, k is assumed to be the same for all items as assumed by previous studies (Hausman et al. 1976).

Consistent with the studies on the required storage-sharing effect in Chap. 2, the realistic RSS for storing item i in class k, in which N_k items share the same region, should be as follows:

$$a_i(N_k) = 0.5\left(1 + N_k^{-\varepsilon}\right) Q(i), \quad 0 < \varepsilon \leq 1. \quad (3.6)$$

ε is the space-sharing factor, and it may be influenced by the initial inventory levels of the items, replenishment policy, ABC demand curve shape, and inventory cost. Fortunately, according to the results obtained in Chap. 2, the value of ε is quite insensitive to these parameters. As a result, $\varepsilon = 0.22$ is adopted in this chapter.

Therefore, under an ABC demand curve with shape factor s, the RSS for class k used to store the N_k items is as follows:

$$R_k = \sum_{i=i_{k-1}+1}^{i_k} a_i(N_k) = \sqrt{0.5KA}\left(1 + N_k^{-\varepsilon}\right) \sum_{i=i_{k-1}+1}^{i_k} \sqrt{(i/N)^s - ((i-1)/N)^s}. \quad (3.7)$$

Furthermore, the cumulative RSS for the first k classes is the sum of the required space of each class:

$$L_k = \sum_{l=1}^{k} R_l = \sqrt{0.5KA}\sum_{l=1}^{k}\left(\left(1 + N_l^{-\varepsilon}\right) \sum_{i=i_{l-1}+1}^{i_l} \sqrt{(i/N)^s - ((i-1)/N)^s}\right). \quad (3.8)$$

Since the warehouse applies an across-aisle policy for class-based storage, the number of sections needed by the first k classes, y_k (this value is also the theoretical further boundary of class k for the continuous model), can be obtained according to the number of aisles in the warehouse,

$$y_k = L_k \big/ (4x + 2), \quad (3.9)$$

where x is the maximum aisle index in the warehouse, and a total of $2x + 1$ aisles exist.

In practice, the class boundaries should be integers, and the practical class boundary for class k, b_k, is obtained as the minimum integer which is no less than y_k:

$$b_k = \lceil y_k \rceil = \lceil L_k / (4x + 2) \rceil . \tag{3.10}$$

Therefore, the storage locations in section b_k may be allocated not only to class k but also to class $k + 1$ ($k < n$) if y_k is not an integer. Here, we further assume that $(y_k + 1 - b_k)(4x + 2)$ storage locations are allocated to class k randomly from section b_k, and the other $(b_k - y_k)(4x + 2)$ are assigned to class $k + 1$.

The warehouse has $2x + 1$ storage aisles, and the farthest aisles from the depot are Aisle x, which are located on the right and left sides of the depot. Thus, the average one-way travel distance from the depot to a location of section b is as follows:

$$d_b = b + \frac{x(x+1)w}{2x+1}. \tag{3.11}$$

Thereafter, the average one-way travel distance for storing or retrieving a unit-load in class k from the depot can be obtained as follows:

$$\begin{aligned} t_k &= \frac{(b_{k-1} - y_{k-1})\, b_{k-1} + \sum_{b=b_{k-1}+1}^{b_k} b - (b_k - y_k)\, b_k}{y_k - y_{k-1}} + \frac{x(x+1)w}{2x+1} \\ &= \frac{y_k b_k - y_{k-1} b_{k-1}}{y_k - y_{k-1}} - \frac{(b_k - b_{k-1})(b_k + b_{k-1} - 1)\, b_k}{2\,(y_k - y_{k-1})} + \frac{x(x+1)w}{2x+1}, \end{aligned} \tag{3.12}$$

where $b_0 = y_0 = 0$. The first part of Eq. (3.12) is the average one-way travel distance in depth, and the second part is that horizontally.

Consequently, according to Eqs. (3.1), (3.4), (3.8), (3.10), and (3.12), the travel distance model of the n-class-based storage system can be obtained as follows:

Model TDM: (Travel Distance Model)

$$\begin{aligned} Min\ \ T_n = &\ \frac{x(x+1)w}{2x+1} \\ &+ \sum_{k=1}^{n} \frac{y_k b_k - y_{k-1} b_{k-1}}{y_k - y_{k-1}} \left(\left(\frac{i_k}{N}\right)^s - \left(\frac{i_{k-1}}{N}\right)^s \right) \\ &- \sum_{k=1}^{n} \frac{(b_k - b_{k-1})(b_k + b_{k-1} - 1)\, b_k}{2\,(y_k - y_{k-1})} \left(\left(\frac{i_k}{N}\right)^s - \left(\frac{i_{k-1}}{N}\right)^s \right), \end{aligned} \tag{3.13}$$

$$s.t.\quad L_k = \sqrt{0.5KA} \sum_{l=1}^{k} \left((1 + N_l^{-\varepsilon}) \sum_{i=i_{l-1}+1}^{i_l} \sqrt{\left(\frac{i}{N}\right)^s - \left(\frac{i-1}{N}\right)^s} \right), \tag{3.14}$$

$$y_k = L_k/(4x+2),\ \ b_k = \lceil y_k \rceil,\ \ k = 1, 2, \cdots, n, \tag{3.15}$$

$$d.v.\quad x,\ i_k,\ k = 1, 2, \cdots, n.$$

For a given number of storage aisles, (i.e., x), the average one-way travel distance for the random policy can be realized when $n = 1$, and that for the full turnover-based policy can be obtained when $n = N$ with only one item in each class. For the class-based policy, the model requires the solution methodology provided in the next section to obtain the optimal number of classes and corresponding class boundaries. Thereafter, according to the solutions for the three policies with different values of x, the optimal layout of the warehouse can be realized, and the performance of the policies can be evaluated.

3.4 Solution Methodology for Class-Based Storage Policy

To find an efficient methodology that obtains the optimal solution for the class-based policy, we first focus on the theoretical continuous model with continuous section index (i.e., $b_k = y_k$ for all the classes) to identify some properties of Model TDM. Thereafter, we decide the optimal number of classes, optimal class boundaries, and minimum average travel time for the realistic (discrete) case. The continuous travel distance model can be written as follows: Model CTM: (Continuous Travel Model)

Model CTM: (Continuous Travel Model)

$$Min \;\; T_n = \frac{x(x+1)w}{2x+1} + \sum_{k=1}^{n} \left(\frac{y_k + y_{k-1}}{2} \left(\left(\frac{i_k}{N}\right)^s - \left(\frac{i_{k-1}}{N}\right)^s \right) \right),$$

$$s.t. \quad \text{Equation (3.14) and } \; y_k = L_k \,/ (4x+2)\,, \quad \text{for } k = 1, 2, \ldots, n,$$

$$d.v. \quad x,\; i_k,\; k = 1, 2, \ldots, n.$$

where $i_0 = 0$, $y_0 = 0$.

From Model CTM, we provide the following theorem, and the corresponding proof can be found in Appendix C.1.

Theorem 3.1 *Denote* $M_n = \sum_{k=1}^{n} \left(\frac{L_k + L_{k-1}}{2} \left(\left(\frac{i_k}{N}\right)^s - \left(\frac{i_{k-1}}{N}\right)^s \right) \right)$, *then for any given number of classes and any ABC demand curve,* $M_n < \sqrt{2NKA}$ *always exists. Consequently, the optimal value of* x *that minimizes the travel distance is* $x^* = \sqrt{(2M_n - w)/(4w)} - 0.5$, *and* x^* *satisfies* $x^* < \sqrt{\sqrt{2NKA}/(2w)}$. *The optimal warehouse shape factor (the ratio of warehouse width,* W, *to warehouse depth,* D) *is* $r_n^* = 2\,(2M_n - w)\,/L_n$ *and satisfies* $r_n^* < 4$.

This theorem shows that the optimal number of aisles in practice is small. Furthermore, the number of aisles is an odd integer in reality. Therefore, we can determine all the optimal classifications of the warehouse for every possible value of x.

We now turn to the practical class boundaries and travel distance model for the realistic discrete case. For a given value of x, Model TDM can be modified as follows:

Model SM: (Solution Methodology)

$$\begin{aligned}
&Min \quad T_n^{'} = \sum_{k=1}^{n} \frac{y_k b_k - y_{k-1} b_{k-1}}{y_k - y_{k-1}} \left(\left(\frac{i_k}{N} \right)^s - \left(\frac{i_{k-1}}{N} \right)^s \right) \\
&\qquad - \sum_{k=1}^{n} \frac{(b_k - b_{k-1})(b_k + b_{k-1} - 1) b_k}{2(y_k - y_{k-1})} \left(\left(\frac{i_k}{N} \right)^s - \left(\frac{i_{k-1}}{N} \right)^s \right), \\
&s.t. \quad \text{Equations (3.14) and (3.15)} \\
&d.v. \quad i_k, \; k = 1, 2, \ldots, n.
\end{aligned} \tag{3.16}$$

For Model SM, the solution of an n-class-based storage system can be found through a dynamic programming algorithm described as follows.

First, k is defined as the index of the stage that corresponds to the kth class. The item with the lowest demand in class k, i_k, denotes the total number of items in the first k classes. i_k is bounded by k and $N-n+k$ because at least one item exists in each class. Second, the number of items for class k, N_k, is the decision variable at stage k, and $1 \leq N_k \leq i_k - k + 1$ because at least $k-1$ items are assigned to the previous $k-1$ classes. Third, the state transfer function of the model is $i_k = N_k + i_{k-1}$, and the evaluation function at stage k for the given i_k is $f_k(i_k)$.

Thereafter, the recursive function can be written as follows according to Eq. (3.16):

$$\begin{aligned}
f_k(i_k) = f_{k-1}(i_{k-1}) + \frac{y_k b_k - y_{k-1} b_{k-1}}{y_k - y_{k-1}} \left(\left(\frac{i_k}{N} \right)^s - \left(\frac{i_{k-1}}{N} \right)^s \right) \\
- \frac{(b_k - b_{k-1})(b_k + b_{k-1} - 1) b_k}{2(y_k - y_{k-1})} \left(\left(\frac{i_k}{N} \right)^s - \left(\frac{i_{k-1}}{N} \right)^s \right),
\end{aligned} \tag{3.17}$$

with initialization $f_0(0) = i_0 = N_0 = 0$, and $f_0(i) = +\infty$ for any $1 \leq i \leq N$.

Thereafter, the minimum objective value at stage $k > 1$ based on a known optimal solution at stage $k-1$ can be found by,

$$\begin{aligned}
f_k^*(i_k) = \min_{1 \leq N_k \leq i_k - k + 1} \Bigg\{ f_{k-1}^*(i_{k-1}) + \frac{y_k b_k - y_{k-1} b_{k-1}}{y_k - y_{k-1}} \left(\left(\frac{i_k}{N} \right)^s - \left(\frac{i_{k-1}}{N} \right)^s \right) \\
- \frac{(b_k - b_{k-1})(b_k + b_{k-1} - 1) b_k}{2(y_k - y_{k-1})} \left(\left(\frac{i_k}{N} \right)^s - \left(\frac{i_{k-1}}{N} \right)^s \right) \Bigg\},
\end{aligned} \tag{3.18}$$

where y_k and b_k are determined according to Eqs. (3.14) and (3.16).

For any given n, $T_n^{'} = f_n^*(i_n = N)$ provides the minimum objective value through the optimization of the number of items in each class (the optimal class boundaries can be found according to the state transfer function $i_k = N_k + i_{k-1}$, and $i_0 = 0$). Moreover, the optimal solution (including the optimal number of classes and corresponding class boundaries) for a given value of x can be obtained through

$n^*(x) = \arg\min_{1 \le n \le N} \{f_n^*(N)\}$. Finally, the optimal item classification and corresponding shortest travel distance for the class-based storage policy can be found according to Eq. (3.13).

Through the dynamic programming algorithm applied for other possible values of x, the optimal number of storage aisles can be found through a comparison of their corresponding minimum travel distances. Thereafter, the optimal warehouse parameters can be obtained.

3.5 Performance Evaluation

In this section, the performance evaluations for the random, class-based, and full turnover-based storage policies are verified numerically based on the proposed travel distance model. The results are based on the following basic parameter values: the aisle width is set to four unit-load widths to provide the forklift with enough space to travel. $N = 100$ items are stored in the warehouse with a total annual demand $A = \sum_{i=1}^{N} D(i) = 10{,}000$. All the items are replenished with the ratio of reorder cost to holding cost $K = 2$, and the space-sharing factor is $\varepsilon = 0.22$ according to the results of Chap. 2. Furthermore, eight ABC demand curves are considered: under each curve, the first 20 % items contribute 20–90 % to the warehouse demand, respectively. Among the curves, the 20 %/90 % curve has the highest skewness, and the 20 %/20 % curve has the lowest skewness because all the items have the same demand volume. The shape of the curves is illustrated in Fig. 3.2.

Through the solution methodology provided in Sect. 3.4, the optimal solutions for the class-based policy can be obtained very fast . The optimal classification under various demand curves are presented in Table 3.2. Thereafter, the one-way travel

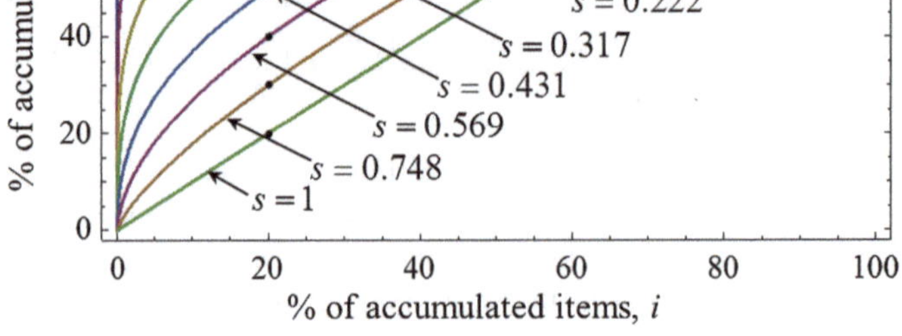

Fig. 3.2 ABC-demand curves

Table 3.2 Optimal classification for class-based storage policy under different demand curves

Shape factor of demand curve s	Optimal number of classes	Optimal number of items in each class (also can be considered as class boundaries)					Total number of items
		Class 1	Class 2	Class 3	Class 4	Class 5	
1 (20 %/20 %)	1	100	–	–	–	–	100
0.748 (20 %/30 %)	2	68	32	–	–	–	100
0.569 (20 %/40 %)	3	22	63	15	–	–	100
0.431 (20 %/50 %)	4	9	43	46	2	–	100
0.317 (20 %/60 %)	4	4	26	52	18	–	100
0.222 (20 %/70 %)	4	2	19	45	34	–	100
0.139 (20 %/80 %)	5	1	11	38	36	14	100
0.065 (20 %/90 %)	5	1	9	25	35	30	100

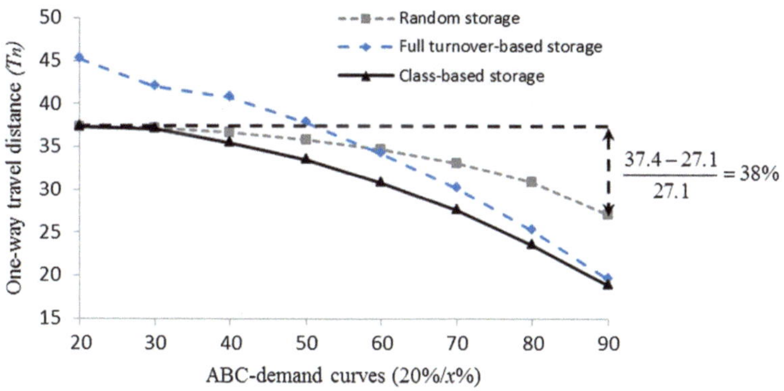

Fig. 3.3 One-way travel distance as a function of demand curves under different storage policies with optimal item classification and optimal warehouse factor

distances for the three polices (random, full turnover, and class-based) under the eight demand curves are presented in Fig. 3.3 as a function of different ABC demand curves, and the corresponding realistic RSSs are shown in Fig. 3.4. Table 3.3 summarizes the optimal warehouse parameters (optimal number of storage aisles in the horizontal and optimal number of sections in depth) under different storage policies and various demand structures, and Fig. 3.5 shows the corresponding warehouse shape factors.

The results in Table 3.2 indicate that a small number of classes yield the optimal solution for the class-based policy. Even for the 20 %/90 % curve with $s = 0.065$, which requires the maximum number of classes for optimal item classification, the optimal number of classes is only 5. In other words, a small number of classes ($n^* \leq 5$) yield the minimum expected one-way travel distance for a unit-load warehouse with a parallel-aisle configuration and an across-aisle storage policy. This result

Table 3.3 Optimal warehouse parameters for different storage policies under various demand curves

Shape factor of demand curve, *s*	Random policy				Full turnover policy				Class-based policy			
	No. of required locations	No. of aisles	No. of sections	Location utilization (%)	No. of required locations	No. of aisles	No. of sections	Location Location (%)	No. of required locations	No. of aisles	No. of sections	Location utilization (%)
1 (20 %/20 %)	1364	19	36	99.71	2000	23	44	98.81	1364	19	36	99.71
0.748 (20 %/30 %)	1350	19	36	98.68	1980	21	48	98.21	1362	19	36	99.56
0.569 (20 %/40 %)	1311	19	35	98.57	1923	21	46	99.53	1342	17	40	98.68
0.431 (20 %/50 %)	1256	17	37	99.84	1842	19	49	98.93	1308	17	39	98.64
0.317 (20 %/60 %)	1175	17	35	98.74	1724	17	51	99.42	1236	15	42	98.10
0.222 (20 %/70 %)	1073	17	32	98.62	1574	15	53	98.99	1141	13	44	99.74
0.139 (20 %/80 %)	929	15	31	99.89	1363	13	53	98.91	1014	11	47	98.07
0.065 (20 %/90 %)	713	13	28	97.94	1046	9	59	98.49	798	9	45	98.52

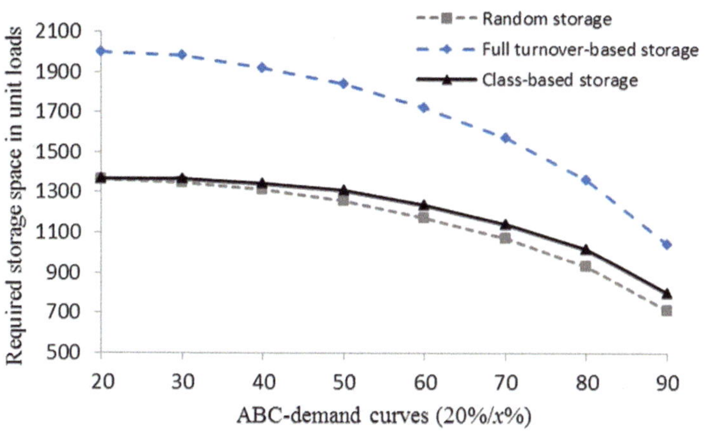

Fig. 3.4 RSS as a function of demand curves under different storage policies with optimal item classification and optimal warehouse factor

is consistent with the findings in Chap. 2 and in practical operations (Roodbergen and Vis 2009).

The results in Fig. 3.3 indicate that the one-way travel distance decreases with the skewness of the demand curve under all three storage policies. An ABC demand curve with high skewness has a small portion of items contributing to the most demand, and so the total RSS for the warehouse being smaller with a classic EOQ replenishment policy (as shown in Fig. 3.4, the RSS decreases under all policies). As a result, the average one-way travel distance is shortened when most portions of the demand are contributed by a small percentage of the stored items. For example, for the random policy, the average one-way travel time (or distance) in conventional research is the same under different demand curves because the RSS of the warehouse is normalized by the total average inventory level. When the realistic RSS is considered, there is almost a 40 % difference between the travel times under the 20 %/20 % curve and the 20 %/90 % curve, and this difference is overlooked in conventional research.

The results also show that the full turnover-based policy performs worse than the class-based policy for all the eight demand curves and performs even worse than the random policy for the first four cases when the skewness of the demand curve is low. In other words, if the items in the warehouse have similar demand volumes, then warehouse managers should not use the full turnover-based policy but should instead consider the random policy. This recommendation is a result of the tradeoff between the space-sharing effect among the items stored in an identical class and turnover ranking according to item demands. The full turnover-based policy has the advantage of turnover ranking in shortening the distance traveled by increasing the weight of retrievals close to the depot but is incapable of space sharing because only one item exists in each class. Consequently, the class-based policy is highly

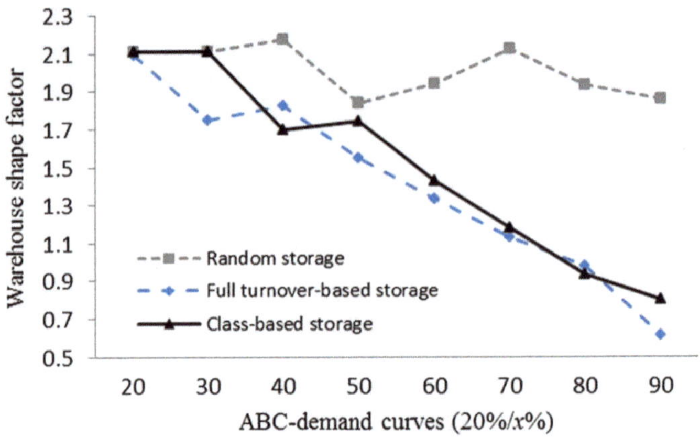

Fig. 3.5 Optimal warehouse shape factor as a function of demand curves under different storage policies

recommended because it balances the tradeoff and takes advantage of the turnover ranking and the space-sharing effect. Although more storage spaces are required for the warehouse under the class-based storage policy than for that under the random policy, the space gap is smaller than that between the full turnover-based policy and the random policy as shown in Fig. 3.4.

Figure 3.5 indicates that the optimal warehouse shape factor is influenced by the demand structure of the stored items and seems to be a decreasing function based on the skewness of the ABC curve under the class-based storage and full turnover-based policies (this case can be considered as a special case of class-based storage with $n = N$). This result suggests that warehouse managers should set a wide-shallow warehouse layout with a large number of short storage aisles for items with similar demand volumes (i.e., flat ABC demand curves with big s) and a narrow-deep warehouse layout with a small number of long storage aisles for items with significantly different demand volumes (i.e., quite skewed ABC demand curves with small s). For example, as shown in Table 3.3, the optimal layout of a warehouse for the 20 %/20 % and 20 %/30 % demand curves is 19 aisles horizontally and 36 sections in depth, whereas that for the 20 %/90 % curve is only 9 aisles horizontally and 45 sections in depth for each storage aisle.

3.6 Chapter Summary

This chapter examines storage policy performance considering realistic RSS for a unit-load warehouse with a given annual demand. Through a generalized travel distance model for an n-class-based storage system, some interesting findings overlooked by existing studies are found. First, the average one-way travel distance for the random policy does not keep constant but decreases with the skewness of the ABC demand curve, and a significant gap (near 40 %) exists between the 20 %/20 % and 20 %/90 % curves. This result indicates that the storage policy performs better with more skewed demand curves.

Second, the results suggest that the full turnover-based policy (a special case of the class-based policy with only one item in each class) will no longer lead to the global minimum average one-way travel distance for the warehouse in contrast to the results of conventional studies that consider the full turnover-based policy as a lower bound of the class-based policy (Hausman et al. 1976; Thonemann and Brandeau 1998). This finding is a result of the fact that the full turnover-based policy requires nearly 50 % extra storage space than the random policy considering the effect of space sharing among items stored in the same class. Consequently, a tradeoff exists between item rankings and space sharing for the number of classes and for the number of items in each class. The result demonstrates that because of reasonably balancing this tradeoff, the class-based policy with a small number of classes (no more than five), but not the full turnover-based policy, is highly recommended for warehouse managers. This finding is consistent with the result in Chap. 2, which examines the class-based storage for a SIT AS/RS with a finite number of items.

In addition, this study also provides the optimal warehouse shape factor (ratio of warehouse width to depth) for different storage policies under various demand curves. The results suggest that warehouse managers should adopt a wide-shallow layout (a large number of short storage aisles) if the stored items have similar demand volumes per unit time (i.e., with a flat ABC demand curve) and a narrow-deep layout (small number of long storage aisles) if the stored items have significantly different demand volumes (i.e., with a quite skewed ABC demand curve). For instance, for the 20 %/40 % demand curve, the optimal warehouse shape factor is 1.7 with 17 storage aisles, whereas for the 20 %/90 % demand curve, the optimal shape factor is 0.8 with only nine storage aisles.

In future studies, this research can be extended in several directions. First, we only considered the across-aisle storage policy for classification although the within-aisle and diagonal policies are also widely applied in practice. Thus, studying the warehouse layout and the optimal classification of these policies will also be useful in practice. Second, research on the optimal warehouse layout based on the flying-V or fishbone aisle configurations considering the space-sharing effect is recommended. Third, as this research focused on a unit-load warehouse operating under a single command, similar research on dual-command operations in an order-picking system can also be interesting. Finally, our results are based on the assumption of having one depot located in the middle of the front aisle. The results from relaxing this assumption may also be useful in practice.

References

Adams ND (1996) Warehouse and distribution automation handbook. McGraw-Hill, Pennsylvania

Ashayeri J, Heuts R, Valkenburg M, Veraart H, Wilhelm M (2002) A geometrical approach to computing expected cycle times for zonebased storage layouts in as/rs. Int J Prod Res 40(17):4467–4483

De Koster R, Le-Duc T, Roodbergen KJ (2007) Design and control of warehouse order picking: a literature review. Eur J Oper Res 182(2):481–501

De Koster RB, Le-Duc T, Zaerpour N (2012) Determining the number of zones in a pick-and-sort order picking system. Int J Prod Res 50(3):757–771

Graves SC, Hausman WH, Schwarz LB (1977) Storage-retrieval interleaving in automatic warehousing systems. Manag Sci 23(9):935–945

Hausman WH, Schwarz LB, Graves SC (1976) Optimal storage assignment in automatic warehousing systems. Manag Sci 22(6):629–638

Jarvis JM, McDowell ED (1991) Optimal product layout in an order picking warehouse. IIE Trans 23(1):93–102

Koh S, Kim B, Kim B (2002) Travel time model for the warehousing system with a tower crane s/r machine. Comput Ind Eng 43(3):495–507

Kouvelis P, Papanicolaou V (1995) Expected travel time and optimal boundary formulas for a two-class-based automated storage/retrieval system. Int J Prod Res 33(10):2889–2905

Le-Duc T, De Koster RB (2005) Travel distance estimation and storage zone optimization in a 2-block class-based storage strategy warehouse. Int J Prod Res 43(17):3561–3581

Petersen CG (2002) Considerations in order picking zone configuration. Int J Oper Prod Manag 22(7):793–805

Petersen CG, Aase G (2004) A comparison of picking, storage, and routing policies in manual order picking. Int J Prod Econ 92(1):11–19

Petersen CG, Schmenner RW (1999) An evaluation of routing and volume-based storage policies in an order picking operation. Decis Sci 30(2):481–501

Petersen CG, Aase GR, Heiser DR (2004) Improving order-picking performance through the implementation of class-based storage. Int J Phys Distrib Logist Manag 34(7):534–544

Rao SS, Adil GK (2013) Optimal class boundaries, number of aisles, and pick list size for low-level order picking systems. IIE Trans 45(12):1309–1321

Roodbergen KJ (2012) Storage assignment for order picking in multiple-block warehouses. In: Warehousing in the Global Supply Chain. Springer, Heidelberg, pp 139–155

Roodbergen KJ, Vis IF (2009) A survey of literature on automated storage and retrieval systems. Eur J Oper Res 194(2):343–362

Rosenblatt MJ, Eynan A (1989) Notederiving the optimal boundaries for class-based automatic storage/retrieval systems. Manag Sci 35(12):1519–1524

Teunter RH, Babai MZ, Syntetos AA (2010) Abc classification: service levels and inventory costs. Prod Oper Manag 19(3):343–352

Thonemann UW, Brandeau ML (1998) Note. optimal storage assignment policies for automated storage and retrieval systems with stochastic demands. Manag Sci 44(1):142–148

Tompkins JA, White JA, Bozer YA, Tanchoco JMA (2010) Facilities planning. Wiley, Hoboken

Van den Berg J (1996) Class-based storage allocation in a single-command warehouse with space requirement constraints. Int J Ind Eng 3:21–28

Yu Y, de Koster RB (2009) Optimal zone boundaries for two-class-based compact three-dimensional automated storage and retrieval systems. IIE Trans 41(3):194–208

Yu Y, Koster R, Guo X (2015) Class-based storage with a finite number of items: using more classes is not always better. Prod Oper Manag 24(8):1235–1247

Zaerpour N, de Koster RB, Yu Y (2013) Storage policies and optimal shape of a storage system. Int J Prod Res 51(23–24):6891–6899

Part II
Optimal Maintenance Support Strategies for Warehousing Equipment

The previous part of this book studies the optimal storage policies of warehousing systems. The response efficiency of warehousing systems depends not only on storage policies but also on the availability of the S/R machine (warehousing equipment). For example, an S/R machine is composed of engines, a track in the horizontal direction, an elevator in the vertical direction, and conveyer systems in depth. The S/R machine cannot work if any one of the components is broken. This constraint makes the maintenance support of the equipment a critical issue in warehouse management. Consequently, the second part of this book focuses on optimal maintenance strategies subject to maximum expected profit of storage enterprise. In Chap. 4, the optimization problem of the storage enterprise is discussed to determine the optimal spare inventory and cost reduction effort when the enterprise performs equipment maintenance on its own. Different from the literature, this study centers on the maximal profit of the enterprise considering the operational requirement of the S/R machine instead of a given availability requirement. In addition, this study extends the model to an alliance model by considering multiple enterprises in the same market. Based on the analysis in Chap. 4, Chap. 5 examines the maintenance strategy in a decentralized scenario when the support service is provided by the maintenance suppliers. A Stackelberg game model is adopted to describe the relationship between the storage enterprises and the maintenance suppliers according to a performance-based contract.

Chapter 4
Optimal Maintenance Decisions in a Self-Maintenance Scenario

Abstract In consideration of the importance of equipment maintenance in warehouse operations, this chapter examines optimal maintenance decisions, that is, the optimal inventory level of spare parts and the optimal effort level in reducing the cost, in a self-maintenance scenario: the storage enterprise performs equipment maintenance on its own. An optimization model that considers the profitability of the equipment is built to maximize the enterprise's expected profit. Results suggest that the optimal inventory level for spare parts increases with the profitability of the equipment, as a result, the optimal required availability of the equipment also increases with profitability. The enterprise incurs a significant profit loss if it seeks a high equipment availability target regardless of its corresponding profitability in the market. In addition, an extended model considering multiple enterprises in the same market is analyzed. Results indicate that pooling the inventory is only appropriate when the unit cost of the spare parts is low. However, if the unit cost is higher than a threshold, then separating the inventories according to enterprise is better than setting a shared inventory through an established alliance.

4.1 Research Background

To maintain a high operational performance of the S/R machines, a storage enterprise sets spare inventories for corresponding critical components (subsystems) such as engines, tracks, and elevators. Practical application and theoretical research indicate that the management of spare inventory is different from that of traditional products (including finished and semi-finished products) (Kennedy et al. 2002; Wang 2012). As indicated by Kennedy et al. (2002), the critical difference between spare inventory and traditional inventory is that spare inventory is not established to fulfill customer demand and is affected by maintenance strategy (e.g., the part is changeable or repairable). Furthermore, spare inventory is affected by the reliability of the equipment, which is unknown to managers in advance. The stockout of spare inventory interrupts the operation of the enterprise and incurs immeasurable loss from economic and reputational perspectives.

X. Guo, *Storage Policies and Maintenance Support Strategies in Warehousing Systems*, Springer Theses, DOI 10.1007/978-981-10-1448-2_4

Among existing literature, Nahmias (1981), Rustenburg et al. (2001), and Kennedy et al. (2002) provided detailed reviews on spare part inventory management and future research directions. In their studies, the topic of the optimization modeling of spare inventory received considerable attention from scholars. Petrović et al. (1990) indicated that the optimization of spare inventory level depends on the analysis of information and conventional data. Haneveld and Teunter (1997) proposed an algorithm that could provide the near-optimal inventory for expensive and durable spare parts. Considering the innovation and competition among different companies, Cohen et al. (1986) found that the strategy showing that the optimal inventory level based on balancing the holding cost and transportation cost restricts enterprises' development. For other related studies on the optimization modeling of spare inventory, see Cohen et al. (1992); Gajpal et al. (1994) and others.

The critical components of assembled equipment are durable in practice, and breakdown is mostly caused by the damage of small parts or software interrupts. In accordance with the existing literature, this book assumes that the critical components are repairable and that the repaired broken components can be considered as new ones (Allen and D'Esopo 1968; Silver 1972; Simpson 1978; Gross and Ince 1978; Gross and Pinkus 1979; Gross 1981; Nakagawa 1981; Kohlas and Pasquier 1981; Yeralan et al. 1986; Brammer and Malmborg 1987; Gupta and Albright 1992; Dhakar et al. 1994; Alfredsson 1997; Wang 2011, 2012).

However, most of the existing studies aim to determine the optimal decision leading to the minimal maintenance cost under a given availability target, whereas in practice, the availability requirement of the equipment depends on its profitability in the market. For example, the availability requirement is small for storage systems in the off-season facing low demand. Furthermore, seeking a high availability of S/R machines in this situation incurs significant profit loss for the enterprise.

To fill the gap between the existing literature and practice, this chapter examines the optimal spare inventory by considering the operational requirement of the enterprise (i.e., the profit produced by the machine and hereafter defined as "profitability"). The effects of profitability on the optimal spare inventory level, realized availability of the equipment, and profit of the enterprise are analyzed. The following section describes the research problem and its corresponding model.

4.2 Problem Description and Modeling

We consider a storage enterprise with N identical assembled S/R machines (we also use systems or equipment in this book). Each one of the N assembled systems with profitability P is composed of n separate critical major components (also called subsystems and denoted as $i = 1, 2, \ldots, n$). Each one of the n major subsystems broken will interrupt the operation of the machine and incur a loss of P. To meet the required availability of the corresponding part of the assembled system, the enterprise holds a spare inventory s_i and a repair facility with infinite capacity to provide repair service with a $M/G/\infty$ queue for subsystem i.

Suppose that the failure of subsystem i occurs at a Poisson rate, λ_i, and is independent of the failures of other subsystems. The broken subsystem is immediately replaced by a working one if spares are available in the spare inventory. Otherwise, a backorder occurs if the corresponding inventory is zero and the replacement is unavailable, and this backorder causes profit loss for the enterprise. The number of backorders for subsystem i is denoted as B_i. The average repair lead time for each failed part is L_i, which includes two parts: the average transportation, disassembly, and setup time l_i from the spare inventory to the corresponding repair facility and the repair time for the part $(L_i - l_i)$. After the failed part is repaired, it is transferred to the spare inventory as a replacement for the next failed part.

Consequently, the demand for each part, denoted as O_i, is the number of failed items in one repair cycle, which is the unit time. According to the assumptions presented above, O_i is a Poisson-distributed variable with mean $\mu_i = \lambda_i L_i$. Furthermore, without loss of generality, the Poisson distribution of O_i can be normalized to a normal distribution with the same mean value and variance in this model (Feeney and Sherbrooke 1966; Kim et al. 2007; Ng et al. 2008). The cumulative distribution function (cdf) and probability density function (pdf) of the distribution are represented by F_i and f_i, respectively, and both have nonnegative support $[0, \infty)$. Thus, the distribution of $B_i = (O_i - s_i)^+$ can be obtained on the basis of O_i and its corresponding inventory level s_i with the mean $E[B_i|s_i] = \int_{s_i}^{+\infty} (1 - F_i(x))\, dx$.

Moreover, when a subsystem fails, a disutility of the system occurs because of the average transportation, disassembly, and setup time, l_i. In unifying the disutility with the enterprise's profit, the expected disutility of subsystem i can be represented by a backorder value represented as $E[b_i] = \lambda_i l_i = \mu_i l_i / L_i$.

Based on the description presented so far, the availability of subsystem i is $A_i = 1 - (B_i + b_i)/N$. The failure of a subsystem in the system is independent of the failure of other subsystems, and the probability that two or more subsystems of the same system failing at any point in time can be negligible. Therefore, a common assumption in the literature (Kim et al. 2007; Muckstadt 2005) states that the availability of the assembled system with n major subsystems is as follows:

$$A_0 = 1 - \frac{\sum_{i=1}^{n} (B_i + b_i)}{N}. \tag{4.1}$$

Furthermore, the expected revenue of the enterprise provided by each assembled system is a linear function of system availability with a coefficient P, which is the revenue provided by the available system to the enterprise per unit time. Consequently, for an enterprise with N identical systems, the expected revenue is as follows:

$$R(A_0) = P\left(N - \sum_{i=1}^{n} (B_i + b_i)\right). \tag{4.2}$$

Lemma 4.1 *The expected revenue that an enterprise obtains from a system increases linearly with the profitability and availability of the system.*

This lemma can be realized from the first-order derivatives of Eq. (4.2). In maintaining a reasonable availability of the system, the total cost of the enterprise for maintaining subsystem i, C_i, consists of fixed and variable components. The fixed component has a mean value of ε_i and can be reduced by an amount a_i via a cost reduction effort provided by the enterprise. The effort costs the enterprise an investment $\psi_i(a_i)$, which increases convexly, and $\psi_i(0) = 0$. We further assume a quadratic function $\psi_i(a_i) = k_i a_i^2/2$ with $k_i > 0$, which has been widely used in scientific research, to express the investment (Chen 2005; Kim et al. 2007). Moreover, without influencing our model, we can normalize the mean of the fixed component to zero. The unit variable cost for each spare part is denoted by c_i; thus, the variable cost of subsystem i is $c_i s_i$, which is the unit cost times the number of spare parts. Consequently, the enterprise's expense in maintaining the subsystem i is $C_i = c_i s_i - a_i$, with an investment $\psi_i(a_i) = k_i a_i^2/2$ made for the cost reduction effort.

With the expected revenue of the enterprise shown in Eq. (4.2), the expected profit of the enterprise can be obtained as follows:

$$\pi_0 = P\left(N - \sum_{i=1}^{n}(B_i + b_i)\right) - \sum_{i=1}^{n}\left(c_i s_i - a_i + \frac{k_i a_i^2}{2}\right). \tag{4.3}$$

The first part is the revenue of the enterprise derived from the performance of the systems, and the second part is the net cost of maintenance. The following subsection analyzes the optimal decisions of the enterprise.

4.3 Optimal Spare Inventory

This subsection first analyzes the optimal spare inventory decisions for each subsystem and then illustrates the optimal decisions and expected profits numerically.

4.3.1 Decision Analysis

The decision variables of the enterprise are the optimal inventory levels, s_i, which are established for spare parts to provide the required availability of the system, and the optimal cost reduction effort, a_i, which is determined to reduce the maintenance cost. Therefore, the problem is as follows:

$$\begin{aligned} &max \quad \pi_0 = P\left(N - \sum_{i=1}^{n}(B_i + b_i)\right) - \sum_{i=1}^{n}\left(c_i s_i - a_i + \frac{k_i a_i^2}{2}\right), \\ &d.v. \quad a_i,\ s_i \geq 0, \quad i = 1, 2, \ldots, n. \end{aligned}$$

The first-best solution of this problem is presented in Proposition 4.1.

Proposition 4.1 *The optimal decisions of the enterprise in obtaining the maximum expected profit considering the profitability of the system are as follows:*

$$s_i^{FB} = F_i^{-1}\left[\max\{1 - c_i\,/P\ ,\ F_i(0)\}\right], \tag{4.4}$$

$$a_i^{FB} = 1/k_i. \tag{4.5}$$

On the basis of the optimal decisions presented in Proposition 4.1, we find that the optimal spare inventory levels shown in Eq. (4.4) depend not only on the unit cost of the corresponding part but also strongly on the profitability of the system. High inventory levels are adopted for systems with high profitability, and a high inventory is established for parts with a low unit cost in the same system with a unique profitability. This setting is reasonable in practice. If the system has a high average profitability, then the inoperability of the system causes considerable loss for the enterprise and a high spare inventory is adopted. In addition, a low unit cost implies that a low marginal cost is needed to increase the availability of the system; thus, the enterprise establishes a higher inventory level for spare parts with low unit cost. This result is also consistent with actual practice: in general, the unit cost reflects the reliability of the part, such that a part with a high unit cost is more durable and exhibits a lower failure rate than a part with a lower unit cost. Consequently, a part with a higher unit cost can have a lower inventory.

Simultaneously, the optimal cost reduction effort shown in Eq. (4.5) continues to be determined by the coefficient between effort and expense. When the effort efficiency is high and k_i is small, the enterprise focuses more on the cost reduction effort, and the effort level is lower with a large k_i.

With the optimal solution presented in Proposition 4.1, the total expected profit of the enterprise is $P\left(N - \sum_{i=1}^{n}(B_i + b_i)\right) - \sum_{i=1}^{n}\left(c_i s_i^{FB} - 1/(2k_i)\right)$, which consists of two parts. The first part is the revenue obtained from the profitability and availability of the systems, and the second part is the maintenance cost for the systems' required availability. Furthermore, the cost is composed of two terms. The first term is the cost to the enterprise's inventory of owing s_i^{FB} spares, and the second term is the net cost savings via the cost reduction effort. System availability and enterprise profit rely heavily on the profitability of the systems, compelling the adoption of different inventory levels in different periods of market demand.

Corollary 4.1 *Based on Eqs. (4.1) and (4.4), the realized optimal availability of the systems under the spare inventory levels presented in Proposition 4.1 increases with the systems' average profitability in the market.*

Proposition 4.2 *The expected profit of the enterprise increases with the average profitability of the systems in the market.*

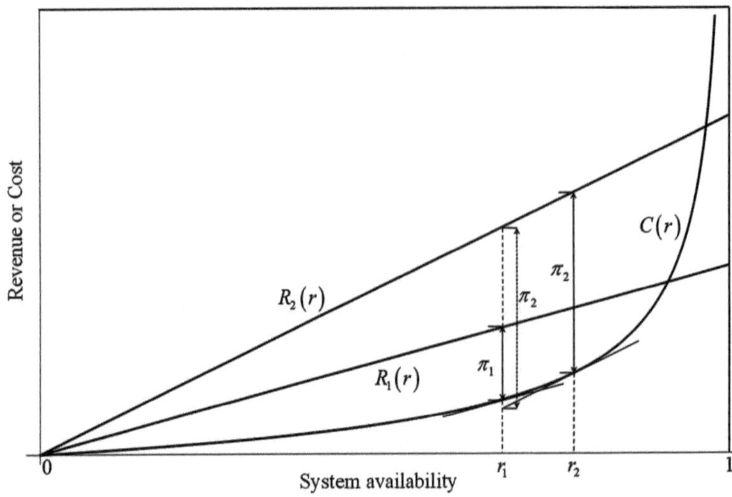

Fig. 4.1 Optimal profits of the enterprise with different profitabilities

Proof According to the analysis in Sect. 4.2, the maintenance cost for keeping the availability of subsystem i at A_i is as follows: $MC_i(A_i) = c_i s_i(A_i) - a_i + \frac{k_i a_i^2}{2}$, where $s_i(A_i) = \arg\left\{(1 - A_i)N = \int_{s_i}^{\infty}(1 - F_i(x_i))\,dx\right\}$. Therefore, as a result of $\frac{\partial MC_i(A_i)}{\partial A_i} = c_i \frac{\partial s_i(A_i)}{\partial A_i} = \frac{Nc_i}{1-F_i(s_i)} > 0$ and $\frac{\partial^2 MC_i(A_i)}{\partial A_i^2} = \frac{Nc_i f_i(s_i)}{(1-F_i(s_i))^2}\frac{\partial s_i(A_i)}{\partial A_i} = \frac{N^2 c_i f_i(s_i)}{(1-F_i(s_i))^3} > 0$, the maintenance cost increases convexly with the availability target of the subsystem. Thus, on the one hand, the maintenance cost for the assembled system, $MC(A_0) = \sum_{i=1}^{n} MC_i(A_{i0})$, increases convexly with the availability of the system.

On the other hand, the profit of the enterprise is determined by the revenue income and maintenance cost of the system, $\pi(A_0) = \sum_{j=1}^{m} R_j(A_0) - MC(A_0)$.

With the result of Lemma 4.1 stating that the revenue of the integrated supply increases linearly with the availability of the system, the second-order derivative of the profit with respect to the availability of the system is $\frac{\partial^2 \pi}{\partial A_0^2} < 0$ as shown in Fig. 4.1. Thus, according to the first-order condition, the optimal availability that leads to the maximal profit of the enterprise can be obtained. For convenience, we denote the optimal value of the system availability as $A_0^*(P)$ for different profitabilities of the system. Suppose that the optimal value for P_1 is A_1^*, and the corresponding profit of the enterprise is π_1^*. Therefore, for $P_2 > P_1$, we know that $\pi_2\left(A_1^*\right) > \pi_1^*$ because the revenue gained from the system increases linearly with the profitability of the system. Consequently, we obtain $\pi_2^*\left(A_2^*\right) \geq \pi_2\left(A_1^*\right) > \pi_1^*$, where A_2^* is the optimal availability of the system with profitability P_2.

Proposition 4.3 *We denote A^* as the optimal resulting availability of the system with respect to its profitability. Then, (i) when the given availability requirement is higher than A^*, a higher requirement produces a lower expected profit for the*

enterprise, and (ii) the harm caused by the higher requirement is more significant than that caused by the lower requirement condition.

Proposition 4.3 can be realized similarly with Proposition 4.2. This finding indicates that the enterprise making decisions with respect to maximizing profit considering profitability is more reasonable than focusing on minimizing the maintenance costs of performing at a prescribed availability target.

4.3.2 Numerical Illustration

To demonstrate the effect of system profitability on optimal spare inventory, availability requirement, and expected profit clearly, this subsection shows the results numerically. By employing the data used by Kim et al. (2007) and supposing $N = 156$ systems held by the enterprise, five key subsystems can be denoted for each system denoted as a, b, c, d, e. The values of $\{\mu_i\}$ and $\{c_i\}$ are presented in Table 4.1.

Here, we further suppose that (1) the profitability of each assembled system is $P = 50$; (2) the fixed cost is $50c_i$; (3) the cost reduction effort factor for subsystem i is $k_i = 1/(10c_i)|$; and (4) the average transportation, disassembly, and setup time is 1/25 of the total repair time, that is, $L_i = 25l_i$.

Based on the parameters, Fig. 4.2 shows the improvement of the profit with varying profitabilities compared with that with a given availability requirement. The result illustrates that the improvement is significant and can be greater than 400 % when the profitability of the system is small while the given availability is high. We choose two

Table 4.1 Average demand information and unit cost for each subsystem (data are collected from Kim et al. 2007)

Subsystem	a	b	c	d	e
μ_i	10.46	19.36	13.72	16.87	8.43
c_i (in $1000)	21.52	6.60	31.08	8.52	14.85

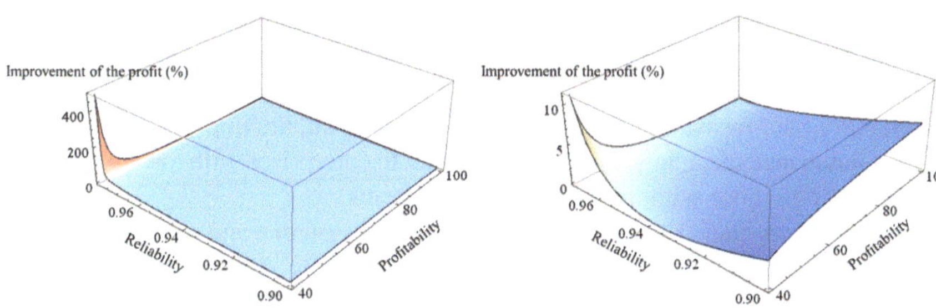

Fig. 4.2 Profit improvement considering system profitability. **a** The profit improvement of PPBC. **b** Close-up view of part *a*

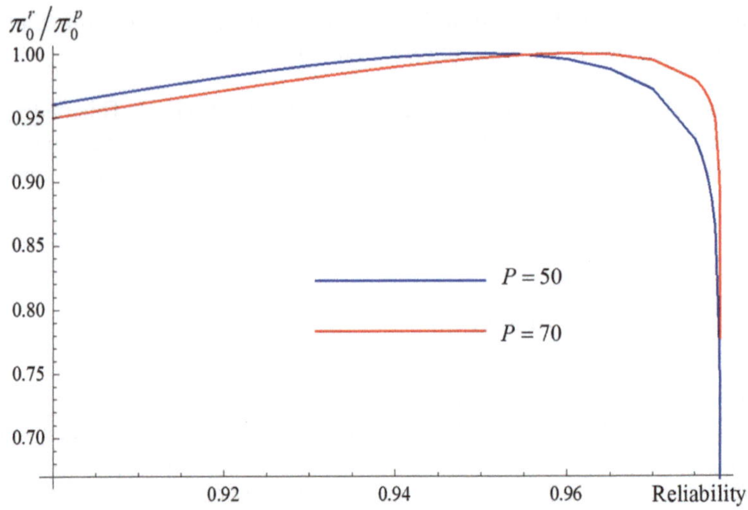

Fig. 4.3 Ratio of the profits with and without a given availability target

cases with different profitabilities of the system ($P = 50$ and $P = 70$, respectively) to describe the importance of the availability choice for the systems in Fig. 4.3.

Figure 4.3 presents the ratio of the profits, π_0^r/π_0^p, gained from the systems under different situations. In the relation, π_0^r is the maximal profit of the enterprise with a given availability target, and π_0^p is the maximal profit of the enterprise with respect to the known profitability of the systems. This figure shows the maximal value of the ratio for both cases, and the maximal value is not greater than 1. In other words, the integrated profit of the enterprise is a convex function with respect to the availability target, and the optimal objective is precisely obtained when the profitability of the system is considered. Additionally, if the availability target is higher than the optimal one, then the profit decreases significantly. In other words, if an enterprise is so selfish in seeking a high performance level regardless of the profitability of the system, then the profit will decrease convexly with the given availability because of the expensive maintenance cost. Moreover, the harm of the wrong choice of the availability requirement is more obvious when the profitability of the system is small (as Fig. 4.2 shows). That is, when the profitability is small, the expense of a high performance level is more memorable than the loss of profit with high profitability but low availability requirement. As a result, the implementation of the right maintenance mechanism is more important to enterprises with low-profitability systems than to those with high-profitability systems.

Figure 4.4 considers two subsystems, namely, subsystem a and subsystem b, with unit cost $c_a < c_b$ and two cases of profitability, namely, case 1 and case 2, with $P_1 < P_2$. The optimal availability of the two subsystems under the two cases of profitability are denoted as r_j^k, where the subscript $j = a$ or b denotes subsystem a

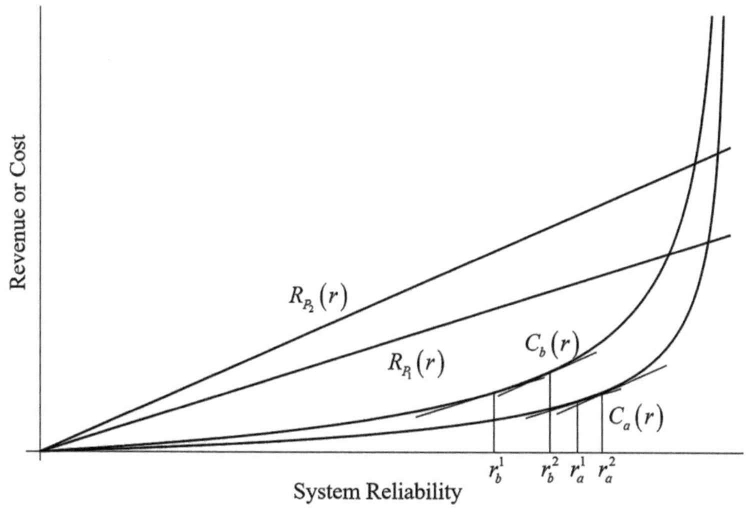

Fig. 4.4 Optimal availability for different subsystems with different profitabilities

or b, and the superscript $k = 1$ or 2 represents the cases with different profitabilities. The following property exists:

Property 4.1 $r_a^k > r_b^k$, *and* $r_j^1 < r_j^2$, *for* $k = 1, 2,\ j = a, b$.

This property indicates that for a certain profitability, a unique optimal availability exists for each subsystem and the system as well. Seeking a high performance level is harmful to enterprises with low-profitability systems because of the expensive maintenance cost for high availability. In addition, the enterprise assumes a significantly higher cost for subsystems with a high unit cost than those with a low unit cost to sustain a certain identical availability of the assembled system. This assumption causes the spare inventory level of each subsystem to decrease with the unit cost.

4.4 Extension: An Alliance Model with Multiple Enterprises

This section considers a scenario with multiple enterprises in the same region or under the supervision of the same parent company, which allows establishing shared inventories of the major components. It verifies the pooling effect of shared inventories among enterprises. Based on the analysis in Sect. 4.2, this section considers m enterprises (represented by the subscript $j = 1, 2, \ldots, m$) and customer j having N_j identically assembled systems with profitability P_j. Therefore, $N = \sum_{j=1}^{m} N_j$ systems exist for the alliance of the m enterprises.

Similar to Sect. 4.3, Proposition 4.4 provides optimal decisions on the spare inventory level and cost reduction effort for the alliance.

Proposition 4.4 *We denote* $\bar{P} = \sum_{j=1}^{m} P_j N_j \Big/ \sum_{j=1}^{m} N_j$ *as the average profitability of all assembled systems in the alliance. Then the optimal decisions of the alliance considering the profitability of the systems are as follows:*

$$s_i^{PI} = F_i^{-1}\left[\max\{1 - c_i / \bar{P}\ ,\ F_i(0)\}\right],$$

$$a_i^{PI} = 1/k_i.$$

Similarly with the findings in Sect. 4.3, Proposition 4.4 indicates that the optimal spare inventory is influenced by the average profitability of the systems and the unit cost of the spare parts. Moreover, the optimal cost reduction effort of the alliance is the same as that in Sect. 4.3 and is not influenced by the alliance.

Proposition 4.5 *When shared spare inventories are established for the alliance of enterprises with different profitability levels, the availability of enterprises with low profitability increases compared with the dedicated inventory scenario. These enterprises are free riders that enjoy the same maintenance support level as that of those with high profitability.*

According to Corollary 4.1, in the dedicated inventory scenario without the alliance, the optimal availability of customers with a low-profitability level is lower than the optimal availability of customers with a high-profitability level. However, when a shared inventory is established in the scenario with alliance, all customers have the same system availability based on the average profitability of the systems. In other words, the system availability for customers with a low profitability is pulled up by customers with a high profitability.

Proposition 4.6 *From the perspective of optimal spare part inventory levels,* $N_j = N_k$ *and* $P_j = P_k$ *for any* j *and* k *in the scenario with homogeneous enterprises. (i) If* $c_i < \bar{P}/2$*, then a pooling effect occurs for the alliance to establish a shared spare part inventory for subsystem* i*. (ii) If* $c_i \geq \bar{P}/2$*, then shared spare inventory increases the cost of the alliance, and maintaining dedicated spare part inventories for each enterprise is preferred.*

Proof of this proposition is presented in Appendix C.2. The finding of Proposition 4.6 contradicts the commonsense view that a shared inventory is always preferable. This finding reflects the high-margin maintenance cost of a specific subsystem. In particular, for spare parts with a low unit cost, establishing a shared inventory to provide maintenance service to the alliance is preferred. For spare parts with an extremely high unit cost, establishing dedicated inventories for each enterprise, such as consignment inventory management for the subsystem in practice, is preferred. Furthermore, on the basis of this proposition, we suggest that for heterogeneous customers with different numbers of systems and different profitability levels, a unique unit cost value, c_i^*, should produce a pooling effect only when the unit cost of a subsystem is lower than c_i^*.

4.5 Chapter Summary

This chapter examines the manner in which an enterprise should determine its optimal spare inventory when it performs the maintenance of its equipment on its own. Traditional studies on spare inventory have mostly focused on the minimization of maintenance cost with a given availability target, disregarding the influence of the systems' profitability in the market. To fill this gap, this chapter analyzes optimal decisions on the inventory level of spare parts by considering the profitability of the systems. Without a given availability target, the enterprise makes decisions on the maximum expected profit by considering the influence of profitability.

The results indicate that the optimal spare inventory of the enterprise significantly depends on profitability. The optimal inventory level is low when the profitability of the system is low. As a result, the optimal required availability of the system is influenced by its profitability in the market. The results of this chapter indicate that the profit of the enterprise decreases if a different (especially a higher) availability target is set regardless of the profitability of the system, and the profit loss can be more than 80 % when the profitability of the system is low.

The analysis on the alliance of multiple enterprises shows that small enterprises are free riders when shared spare inventories are established. Moreover, the pooling effect does not always exist. This finding suggests that a pooling effect exists for components with a low unit cost, and setting a shared spare inventory for the alliance is suitable. However, the pooling effect disappears when the unit cost of the spare part is high because the total spare inventory level for the alliance is higher than the sum of the dedicated inventories of each enterprise. As a result, setting dedicated spare inventories for each enterprise, such as consignment inventory management, is suitable because of the increase of maintenance costs.

In addition, this chapter discusses the optimal decision on the cost reduction effort. The results, which are consistent with those of the existing literature, find that the optimal decision is not influenced by the profitability of the system (Kim et al. 2007).

This chapter reveals that the method adopted in existing studies (i.e., making decisions based on a given availability requirement) may not result in the maximum profit for the enterprise. Thus, research on spare inventory established by maximizing the expected profit, relaxing the constraint of availability target, and considering the profitability of the systems leads to new findings for research and new managerial insights for practice. Chapter 5 examines the optimal maintenance support strategy in an after-sales service supply chain, in which the maintenance service is provided by the original equipment manufacturer.

References

Alfredsson P (1997) Optimization of multi-echelon repairable item inventory systems with simultaneous location of repair facilities. Eur J Op Res 99(3):584–595

Allen SG, D'Esopo DA (1968) An ordering policy for repairable stock items. Op Res 16(3):669–674

Brammer K, Malmborg C (1987) A transient state model for predicting maintenance requirements. Eng Costs Prod Econ 11(1):87–98

Chen F (2005) Salesforce incentives, market information, and production/inventory planning. Manag Sci 51(1):60–75

Cohen MA, Kleindorfer PR, Lee HL (1986) Optimal stocking policies for low usage items in multi-echelon inventory systems. Nav Res Logist Q 33(1):17–38

Cohen MA, Kleindorfer PR, Lee HL, Pyke DF (1992) Multi-item service constrained (s, s) policies for spare parts logistics systems. Nav Res Logist 39:561–577

Dhakar TS, Schmidt CP, Miller DM (1994) Base stock level determination for high cost low demand critical repairable spares. Comput Op Res 21(4):411–420

Feeney GJ, Sherbrooke CC (1966) The (s-1, s) inventory policy under compound poisson demand. Manag Sci 12(5):391–411

Gajpal PP, Ganesh L, Rajendran C (1994) Criticality analysis of spare parts using the analytic hierarchy process. Int J Prod Econ 35(1):293–297

Gross D (1981) Saspro iia spare and server provisioning model. Comput Op Res 8(3):197–207

Gross D, Ince JF (1978) Spares provisioning for repairable items: cyclic queues in light traffic. AIIE Trans 10(3):307–314

Gross D, Pinkus CE (1979) Designing a support system for repairable items. Comput Op Res 6(2):59–68

Gupta A, Albright SC (1992) Steady-state approximations for a multi-echelon multi-indentured repairable-item inventory system. Eur J Op Res 62(3):340–353

Haneveld WK, Teunter R (1997) Optimal provisioning strategies for slow moving spare parts with small lead times. J Op Res Soc. pp 184–194

Kennedy W, Patterson JW, Fredendall LD (2002) An overview of recent literature on spare parts inventories. Int J Prod Econ 76(2):201–215

Kim SH, Cohen MA, Netessine S (2007) Performance contracting in after-sales service supply chains. Manag Sci 53(12):1843–1858

Kohlas J, Pasquier J (1981) Optimization of spare parts for hierarchicall decomposable systems. Eur J Op Res 8(3):294–300

Muckstadt JA (2005) Analysis and algorithms for service parts supply chains. Springer Science & Business Media, Berlin

Nahmias S (1981) Managing repairable item inventory systems: a review. TIMS Stud Manag Sci 16:253–277

Nakagawa T (1981) A summary of periodic replacement with minimal repair at failure. J Op Res Soc Jpn 24(3):213–227

Ng HKT, Filardo G, Zheng G (2008) Confidence interval estimating procedures for standardized incidence rates. Comput Stat Data Anal 52(7):3501–3516

Petrović D, Petrović R, Senborn A, Vujosević M (1990) A microcomputer expert system for advising on stocks in spare parts inventory systems. Eng Costs Prod Econ 19(1):365–370

Rustenburg W, van Houtum GJ, Zijm W (2001) Spare parts management at complex technology-based organizations: an agenda for research. Int J Prod Econ 71(1):177–193

Silver EA (1972) Inventory allocation among an assembly and its repairable subassemblies. Nav Res Logist Q 19(2):261–280

Simpson VP (1978) Optimum solution structure for a repairable inventory problem. Op Res 26(2):270–281

Wang W (2011) A joint spare part and maintenance inspection optimisation model using the delay-time concept. Reliab Eng Syst Saf 96(11):1535–1541

Wang W (2012) A stochastic model for joint spare parts inventory and planned maintenance optimisation. Eur J Op Res 216(1):127–139

Yeralan S, Dieck AJ, Darwin RF (1986) Economically optimum maintenance, repair and buffering operations in manufacturing systems. Eng Econ 31(4):279–292

Chapter 5
Equipment Maintenance Support Under Performance-Based Contracts

Abstract Chapter 4 analyzed the optimal decisions on equipment maintenance when support is performed by the enterprise on its own. This chapter considers the problem in an after-sales service supply chain when maintenance service is provided by manufacturers. This scenario is a common situation in practice. For example, the manufacturer of the airplane provides the maintenance service of aircrafts. Aftermarket maintenance service supply chains are generally composed of two kinds of players, namely, the clients (the customers/enterprises with the systems) and the suppliers who provide the maintenance service. Accordingly, the service contract becomes important in operating a sustainable partnership between the two players. Among all service contracts, the two most popular ones are the fixed-price contract and the cost-plus contract. Under the fixed-price contract, customers pay the service suppliers a fee for the necessary subsystem and service support, whereas under the cost-plus contract, customers cover the full cost of the suppliers and pay them a premium or commission. In recent years, a novel contract, the performance-based contract, has been proposed. Customers pay suppliers according to the availability of the system according to three contract terms: a fixed payment, a cost reimbursement, and a backorder penalty. Existing research has shown that the performance-based contract is more effective than the cost-plus and fixed-price contracts and minimizes the maintenance cost of customers with a given availability target. This chapter studies how to maximize the profits of customers with varying profitabilities under performance-based contracts in a contract horizon. It models the actions of customers and their service suppliers through a multiple-leader-multiple-follower game model. Different from previous studies, this study indicates that the assembled system's availability should be adjusted flexibly according to its profitability. Furthermore, similar to the findings in Chap. 4, the intuition showing customer behavior aiming at the highest possible availability usually damages its profit heavily. At equilibrium, all customers give the same cost reimbursement to suppliers even if they hold different number of systems with different profitabilities. Furthermore, similar to the findings in the previous chapter, if suppliers provide after-sales service for multiple customers through a shared spare inventory, then a pooling effect only exists when the unit cost of the spare part is significantly lower than the average profitability of the systems. Moreover, we are able to design a contract for the decentralized scenario to achieve the performance of the integrated one even though the new performance-based

X. Guo, *Storage Policies and Maintenance Support Strategies in Warehousing Systems*, Springer Theses, DOI 10.1007/978-981-10-1448-2_5

contract performs at a high level and the coordination contract is not necessary in most cases.

5.1 Research Background

Support and maintenance service in the after-sales market has become an important component of economic development, and this service usually provides a higher profit rate to suppliers than the sales of the original products. For instance, IBM received more than $5 billion in revenue from maintenance and service in 2001 (Cohen et al. 2006a). Furthermore, Accenture determined that after-sales service provided profits of $2 billion among the $9 billion in revenue for General Motors in 2003, a significantly higher ratio than the profits from car sales during the same time period (Dennis and Kambil 2003). Consequently, the move toward a service-based economy has transformed firms into service businesses (Cohen et al. 2006b) particularly in industries that produce complex products and in which the consequences of system downtime are severe (Mirzahosseinian and Piplani 2011). Over the life cycle of most systems, 30 % of the entire cost is estimated to be spent on acquiring the system, and 70 % is spent on system support and maintenance (Berkowitz et al. 2005).

After-sales maintenance service supply chains generally have two types of players: clients (i.e., customers employing the systems) and suppliers who provide maintenance services. A service contract is critical for producing a sustainable partnership between the two types of players. The two most popular service contracts are the fixed-price and cost-plus contracts. Under a fixed-price contract, customers pay a fee to service suppliers for the necessary subsystem and service support. Under a cost-plus contract, customers cover the suppliers' full costs and pay a premium or commission. These two contracts are extensively adopted by capital-intensive industries such as aerospace and defense because of significant uncertainties in costs and repair processes (Kim et al. 2007).

On the basis of the research on aerospace and defense, Kim et al. (2007) introduced a novel approach: the performance-based contract (PBC) based on theory of "power by the hour" or "performance-based logistics" in the commercial airline and defense industries, respectively. This contract has three terms: fixed payment, cost reimbursement, and backorder penalties. This contract seeks to minimize maintenance costs for customers under a given availability target and to enable customers to manage their service suppliers, who provide maintenance support services by setting spare part inventories, with minimal cost to maintain the required level of availability of the system. The cost-plus and fixed-price contracts can be viewed as special cases of PBC. In particular, the minimum maintenance cost increases convexly with the availability target because the marginal cost increases under PBC.

In reality, different customers face different demands in their individual markets and achieve different revenues using the same type of systems. For instance, the annual container throughput volume of the Port of Rotterdam, which is the largest port in Europe, is approximately 12 million twenty-foot equivalent units (TEUs); this volume is only 40 % of the throughput volume of the Port of Shanghai (http://www.hafen-hamburg.de/en/content/container-port-throughput-global-comparison). Thus, a container crane provides different utilities for the two ports. Furthermore, even for the same customer, the utility (e.g., revenue or profit) obtained from the same system fluctuates in different demand seasons. For instance, the throughput volume (in number of TEUs) of the Port of Shanghai in February is 15–30 % lower than its annual average level (http://www.portshanghai.com.cn/jtwbs/webpages/server_teu.html). Other industries also exhibit seasonal effects, such as airlines (Oppermann and Cooper 1999; Chung and Whang 2011) and clothing manufacturing (Jin and Sternquist 2003; Şen 2008). Thus, the profitability of a system changes even for the same customer.

Customers with a more profitable system are willing to pay more in support costs to achieve high availability according to marginal cost and marginal revenue, and a trade-off exists between maintenance costs from system availability and revenue obtained from the available systems. The findings of previous studies considering a given availability constraint are incapacitated facing this trade-off. Consequently, the present study relaxes the constraint of a given availability target to determine ways of maximizing customer profits with respect to the system profitability in a performance-based support situation.

An after-sales service supply chain consisting of multiple customers and multiple suppliers is analyzed in this chapter. Determining a closed-form solution is a complex and difficult problem for customers because of the complicated mathematical formulas required. Fortunately, proving the existence of a decentralized equilibrium of the game, which produces interesting findings, is possible. First, even when customers hold a different number of systems and exhibit different profitability levels, they provide the same cost reimbursement to the supplier at equilibrium. Second, the numerical analysis indicates that under optimal contract terms, customer profit increases significantly with profitability, whereas the increase in supplier profit is lower than that of customer profit. Moreover, although customers are the leaders in this game, those with low profitability systems cannot transfer major profits from the supplier. This result coincides with reality because suppliers perform the same maintenance work regardless of the profitability of the system.

Another counterintuitive but interesting result beyond our expectation indicates that the performance incentives (backorder penalties) for suppliers decrease rather than increase as the profitability of the system increases (or as the unit cost of the subsystem decreases), whereas the cost reimbursement paid by customers increases simultaneously. First, customers prefer to share the cost of spare parts to motivate suppliers to maintain high spare inventories. Second, when the required availability of the system is high, the expected number of backorders is low, and thus the penalties in the maintenance contract have little effect.

Lastly, the numerical analysis in the present study reveals that the contract performs at a high level and that the players do not need a coordination contract in most cases, although coordination contract terms are presented in the present paper.

The example in the numerical analysis in Sect. 5.6 indicates that the coordination improvement in most cases is not greater than 5 % with varying system profitabilities.

5.2 Literature Review

For a long time, the problem of performance, availability, and reliability has been studied by numerous experts. Most studies have been focused on reliability modeling in computer and software-based systems and power industries. For instance, Oikonomou (1993) introduced a calculation method for computer-based system reliability based on incomplete information. Thereafter, many articles about the reliability modeling of computer and network-based systems have been published (Robidoux et al. 2010; Mettas 2000; Kang et al. 2008; Wang and Li 2012; Xing and Levitin 2011; Dai et al. 2008). Since the 1980s, an increasing number of researchers have focused on reliability analysis and development policies of the software industry (Butler and Finelli 1993; Cai et al. 1991; Yamada and Osaki 1985; Zio 2009) and on reliability studies about the power industries (Ahmadi-Khatir et al. 2009).

In recent years, the availability issues in the manufacturing industries have also been given much attention (Gaiardelli et al. 2007). Although different from the reliability issues in the computer-based and software industries, the availability of systems in the manufacturing industries is usually provided by suppliers' maintenance and after-sales support. Consequently, customers adopt contracts to motivate suppliers to provide the desired availability. The resource-based contract and the performance-based contract have been discussed in the literature as two major kinds of contracts that describe the relationship between the customers and maintenance suppliers (Kim et al. 2007, 2010; Oliva and Kallenberg 2003; Roels et al. 2010). In the literature, the performance-based contract emerges as a new service model replacing the resource-based contract in the after-market service supply chains. Mirzahosseinian and Piplani (2011) found that increasing the components' reliability and the efficiency of repair facility is more effective than setting a high spare part inventory level. Adding to the modeling analysis of performance contracting, some researchers have conducted empirical studies on performance contracts. Guajardo et al. (2012) empirically investigated the relationship between product reliability and after-sales contract. Through a proprietary data set provided by Rolls-Royce (a major manufacturer of aircraft engines), two types of maintenance contracts are analyzed: time and resource contracts and PBC. An estimation of their model indicates that product availability is higher by 25–40 % under PBC than that under time and resource contracts when the endogeneity of contract choice is considered.

In addition to the availability of the system, the outcome (profit or revenue) is a more important factor that customers are concerned about when they make decisions in practice. The studies reviewed above have focused on the availability of a product or system and have made decisions with respect to the minimal maintenance cost. However, in practice, the outcome (or profit and revenue) is another critical factor for customers. The outcome-based contract is a contracting model that enables customers to pay only when suppliers have delivered the outcomes instead of activities and tasks (Ng et al. 2009). It focuses on the required outcome instead of performing to the prescribed specifications (Gruneberg et al. 2007; Gustafsson et al. 2010).

The following studies come closest to our work. Kim et al. (2007) studied the PBC model through N identical systems, where each system is composed of n critical subsystems. Customers enter a contract with suppliers, who keep the spare inventory and repair facilities to maintain each subsystem. Customers pay a part of supplier's maintenance cost, which can be reduced by supplier's cost reduction effort as compensation, and customers penalize suppliers depending on the number of backorders of each subsystem. For mission-critical systems with infrequent restoration and recovery, supplier will bear expensive cost for committing the necessary resources for recovery because failures occur infrequently and the resources will be idle most of time. To close this gap, Kim et al. (2010) introduced a performance contracting model for the infrequent restoration and recovery of mission-critical systems. They compared the efficiencies of two widely used contracts based on sample-average downtime and cumulative downtime. Thereafter, the authors extended the work of Kim et al. (2007) to two studies focusing on the interaction between component reliability and spare inventory decisions (Kim et al. 2011), and a two-stage sequential game model consisting of multiple risk-neutral suppliers and a customer (Kim 2011). These three studies complement one another and provide significant managerial insights into customers and suppliers under a performance-based contracting scenario.

By reviewing the related literature, we find that the existing performance-based contracts aim to find the optimal contract terms leading to the minimal maintenance cost under a given availability target. However, in practice, customers aim to obtain the maximal profit instead of the minimal maintenance cost. Therefore, customers wisely prefer to provide suppliers with the contract terms according to the maximal profit instead of a given availability target of the system. So far, no study has yet discussed after-sales maintenance contract analysis according to the maximal profit by considering the profitability of the system. To fill this gap, this chapter examines the optimal contract strategy under performance-based maintenance support contracts. Storage systems are taken as an example to maximize the expected profit of customers considering profitability rather than to minimize the maintenance cost with a given availability target.

5.3 Problem Description and Performance-Based Contracts

We consider an after-sales service supply chain composed of m customers (represented by the subscript $j = 1, 2, \ldots, m$), with customer j having N_j identically assembled systems and n maintenance suppliers (represented by the subscript $i = 1, 2, \ldots, n$). Each of the assembled systems is composed of n separate major subsystems produced and maintained by a unique supplier. Therefore, each supplier faces a maintenance market with $N = \sum_{j=1}^{m} N_j$ subsystems.

To avoid repeated statements, we assume that the maintenance process of the systems, demand information of the spare parts, and maintenance cost are the same as those in Chap. 4. Please see Sect. 4.2 for details. Here, we only provide the key related formulas as follows.

The expected backorders of subsystem i with a given spare inventory s_i is as follows:

$$E[B_i|s_i] = \int_{s_i}^{+\infty} (1 - F_i(x))\, dx.$$

The expected availability of the systems is as follows:

$$A_0 = 1 - \frac{\sum_{i=1}^{n} (B_i + b_i)}{N}. \tag{5.1}$$

Therefore, the expected revenue of customer j with N_j systems is as follows:

$$R_j(A_0) = P_j N_j \left(1 - \frac{\sum_{i=1}^{n} (B_i + b_i)}{N}\right).$$

The maintenance cost for supplier i to maintain the corresponding subsystem is as follows:

$$C_i = c_i s_i - a_i,$$

where a_i is the cost reduced by the cost reduction effort based on an investment, $\psi_i(a_i) = k_i a_i^2/2$.

In the principal agent model of the service supply chain, customer j provides a contract of payment

$$T_{ij} = c_i \mu_{ij} + \alpha_{ij} C_i + \gamma_{ij}(A_i - A_{i0}),$$

to supplier i for the maintenance of part i in its system. This payment is composed of three terms. The first term is a fixed payment that depends on the supplier's unit variable cost and mean demand, where $\mu_{ij} = \sum_{j=1}^{m} N_j \mu_i / N$ is the mean value of the demand distribution. The second term is a ratio of the supplier's cost, which may be a reimbursement to the supplier that requires a high maintenance cost or a payback for the fixed payment from the supplier to the customer when the supplier's realized cost is very low (i.e., α_{ij} can be positive or negative in different scenarios). The last term is an incentive for supplier performance, and A_{i0} is the average performance level of the subsystem in the corresponding industry. Consequently, the third term of the contract should be regarded as a reward to the supplier when performance is higher than A_{i0} and should be viewed as a penalty when performance is lower than A_{i0}. To reflect the relationship between the performance incentive and the profitability of the system, we let $\gamma_{ij} = \rho_{ij} N_j P_j$ with $|\rho_{ij}| < 1$. Furthermore, $(A_i - A_{i0})$ can be rewritten as $(B_{i0} - B_i)/N$ based on the relationship between system performance and spare backorders, where B_{i0} is the corresponding value of the total number of backorders for all m customers based on the average performance level of subsystem i. Overall, the payment of customer j to supplier i for maintenance support is as follows:

$$T_{ij}(C_i, B_i) = c_i\mu_{ij} + \alpha_{ij}C_i + \rho_{ij}N_jP_j(B_{i0} - B_i)/N. \tag{5.2}$$

The contract parameters in the model are $\{\alpha_{ij}, \rho_{ij}\}$, where α_{ij} is an incentive for the supplier to establish spare inventory and cost reduction effort, and ρ_{ij} is an incentive only for the spare inventory to achieve sufficient system performance.

According to the assumptions and analysis presented so far, the expected profit of supplier i under the given contracts $T_{ij}(C_i, B_i)$ is as follows:

$$\begin{aligned}\pi_i &= E\left[\left(\sum_{j=1}^{m} T_{ij}(C_i, B_i) - C_i - \psi_i(a_i)\right)\middle| a_i, s_i\right] \\ &= \sum_{j=1}^{m}\left(c_i\mu_{ij} + \alpha_{ij}(c_is_i - a_i) + \frac{\rho_{ij}P_jN_j\left(B_{i0} - E[B_i|s_i]\right)}{N}\right) \\ &\quad - (c_is_i - a_i) - \frac{k_ia_i^2}{2}.\end{aligned} \tag{5.3}$$

The first term is the expected income from m customers, the second term is the cost of maintaining the corresponding subsystem, and the last term is the expense of the investment from the cost reduction effort. Similarly, when contracts are offered to all suppliers, the expected profit for customer j is as follows:

$$\begin{aligned}\Pi_j &= E\left[R_j(A_0) - \sum_{i=1}^{n} T_{ij}(C_i, B_i)\middle| \{a_i, s_i\}\right] \\ &= P_jN_j\left(1 - \sum_{i=1}^{n}\frac{E[B_i|s_i] + b_i}{N}\right) \\ &\quad - \sum_{i=1}^{n}\left(c_i\mu_{ij} + \alpha_{ij}(c_is_i - a_i) + \frac{\rho_{ij}N_jP_j\left(B_{i0} - E[B_i|s_i]\right)}{N}\right).\end{aligned} \tag{5.4}$$

The first term of Eq. (5.4) is the expected revenue obtained from the availability of the working systems at hand, and the second term is the total payment transferred to suppliers for the maintenance of the systems.

5.4 Decision Analysis

In the after-sales service supply chain, customers first offer suppliers the payment contract in light of the contract terms assumed. Thereafter, suppliers who want to accept the contracts determine the levels of their cost reduction effort and spare part

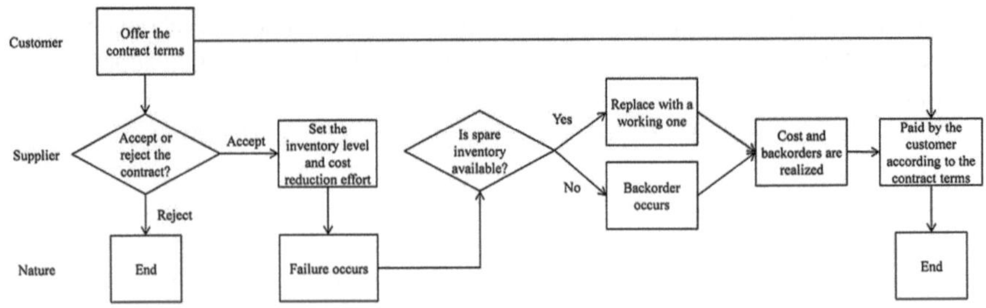

Fig. 5.1 Event sequence in the maintenance support game

inventories. Finally, after the maintenance cost and backorders are realized at the end of the contract horizon, suppliers are paid according to the terms of the contract. Based on the previous analysis, the sequence of events is illustrated in Fig. 5.1.

Under the service contract, customers do not control the details of supplier costs and the methods by which suppliers meet their performance objectives. Therefore, suppliers choose their cost reduction efforts and spare inventories with respect to their own maximal profits, which may not be the optimal choices of customers. Consequently, customers' task in this after-sales support relationship is to employ appropriate incentives through the contract terms to induce suppliers to make decisions that favor the customers. Moreover, the m customers play a Nash game as a multiple-leader in a Stackelberg game with n suppliers.

According to the sequence of events and backward induction, supplier's actions are analyzed first.

5.4.1 *Supplier's Decision*

Given the contract terms $\{\alpha_{ij}, \rho_{ij}\}$, all suppliers face the same problem, that is, determining the level of cost reduction effort and spare inventory by maximizing the expected profit. Therefore, supplier i solves the following:

$$\max_{a_i, s_i} \quad \sum_{j=1}^{m} c_i \mu_{ij} - \left(1 - \sum_{j-1}^{m} \alpha_{ij}\right)(c_i s_i - a_i) + \sum_{j=1}^{m} \frac{\rho_{ij} N_j P_j \left(B_{i0} - E[B_i|s_i]\right)}{N} - \frac{k_i a_i^2}{2}. \tag{5.5}$$

Proposition 5.1 *We denote* $\boldsymbol{\alpha}_i = \sum_{j=1}^{m} \alpha_{ij}$ *and suppose* $\boldsymbol{\alpha}_i < 1$. *A unique optimal solution exists for the supplier's problem, in which* s_i *and* a_i *are*

$$s_i^* = F_i^{-1}\left[\max\left\{1 - N(1 - \boldsymbol{\alpha}_i)c_i \Big/ \left(\sum_{j=1}^{m} \rho_{ij} P_j N_j\right), F_i(0)\right\}\right], \tag{5.6}$$

$$a_i^* = (1 - \boldsymbol{\alpha}_i) \big/ k_i. \tag{5.7}$$

On the basis of the equations of the solution, we know that the optimal a_i^* and s_i^* have the same properties with respect to k_i and P_j as in the first-best solution. Moreover, this solution depends on the contract terms and exhibits further properties that are indicated in Corollary 5.1 as follows:

Corollary 5.1 *(i)* $\partial s_i^* \big/ \partial \alpha_i^* > 0$, $\partial a_i^* \big/ \partial \alpha_i^* < 0$ *and (ii)* $\partial s_i^* \big/ \partial \rho_i^* > 0$, $\partial a_i^* \big/ \partial \rho_i^* = 0$.

This corollary yields findings similar to those of previous studies, such as Kim et al. (2007). In particular, the spare inventory is an increasing function of α_{ij} and ρ_{ij}. The cost reduction effort of supplier i is a decreasing function of the reimbursement ratio α_{ij} and is not affected by the performance incentive ρ_{ij}. Thus, customers provide high performance incentives and reimbursements to suppliers to achieve high system availability. However, a side effect is that the cost reduction effort decreases as reimbursement increases. This decrease in cost reduction effort is due to the increasing marginal effort cost to the supplier caused by the cost reduction shared with the customer through the ratio α_{ij} in contrast to the corresponding expense for effort investments, which are undertaken by the supplier alone.

5.4.2 Customer's Decision

Certifying suppliers' response about the contract terms, the m customers determine the contract terms for each supplier with respect to its maximal profit through a Nash game. Hence, customer j solves the following:

$$\begin{aligned}
Max \quad & \Pi_j = P_j N_j \left(1 - \sum_{i=1}^{n} \frac{E[B_i|s_i] + b_i}{N} \right) \\
& \quad - \sum_{i=1}^{n} \left(c_i \mu_{ij} + \alpha_{ij}(c_i s_i - a_i) + \frac{\rho_{ij} N_j P_j \left(B_{i0} - E[B_i|s_i]\right)}{N} \right), \\
s.t. \quad & s_i = F_i^{-1} \left[\max \left\{ 1 - N(1 - \boldsymbol{\alpha}_i)c_i \Big/ \left(\sum\nolimits_{j=1}^{m} \rho_{ij} P_j N_j \right), F_i(0) \right\} \right], \\
& a_i = (1 - \boldsymbol{\alpha}_i) \big/ k_i, \quad i = 1, 2, \ldots, n, \\
d.v. \quad & \alpha_{ij}, \rho_{ij}.
\end{aligned}$$

With respect to the precondition stating that $P_j N_j$ and b_i are constantly independent from the contract terms, customer j's problem can be split into n identical problems as follows:

$$\begin{aligned} Max \quad & -\alpha_{ij}c_i s_i + \alpha_{ij}(1-\boldsymbol{\alpha}_i)/k_i - \frac{\left(1-\rho_{ij}\right)P_j N_j E[B_i|s_i] + \rho_{ij}P_j N_j B_{i0}}{N}, \\ s.t. \quad & 1-F_i(s_i) = N(1-\boldsymbol{\alpha}_i)c_i \Big/ \left(\textstyle\sum_{j=1}^m \rho_{ij}P_j N_j\right), \\ & \boldsymbol{\alpha}_i = \sum_{j=1}^m \alpha_{ij}, \\ d.v. \quad & \alpha_{ij}, \rho_{ij}. \end{aligned} \tag{5.8}$$

Proposition 5.2 *With the suppliers' responses presented as Eqs. (5.6) and (5.7), although the customer's problem is not generally quasi-convex, the optimal contract terms for the customer must satisfy the following conditions:*

$$\begin{aligned} \alpha_{ij}^{SB} = & \frac{\left(1-\sum_{l=1}^m \alpha_{il}^{SB} - k_i c_i s_i^{SB}\right) f_i\left(s_i^{SB}\right)\sum_{l=1}^m \rho_{il}^{SB}P_l N_l}{f_i\left(s_i^{SB}\right)\sum_{l=1}^m \rho_{il}^{SB}P_l N_l + N k_i c_i^2} \\ & + \frac{\left(1-\rho_{ij}^{SB}\right)\left(1-F_i\left(s_i^{SB}\right)\right)k_i c_i P_j N_j}{f_i\left(s_i^{SB}\right)\sum_{l=1}^m \rho_{il}^{SB}P_l N_l + N k_i c_i^2}, \end{aligned} \tag{5.9}$$

$$\rho_{ij}^{SB} = 1 - \frac{N\alpha_{ij}^{SB}c_i\left(1-F_i\left(s_i^{SB}\right)\right) + \left(B_{i0} - E[B_i|s_i^{SB}]\right) f_i\left(s_i^{SB}\right)\sum_{l=1}^m \rho_{il}^{SB}P_l N_l}{P_j N_j\left(1-F_i\left(s_i^{SB}\right)\right)^2}, \tag{5.10}$$

where $s_i^{SB} = F_i^{-1}\left[\max\left\{1-N\left(1-\sum_{j=1}^m \alpha_{ij}^{SB}\right)c_i \Big/ \left(\sum_{j=1}^m \rho_{ij}^{SB}P_j N_j\right), F_i(0)\right\}\right]$, *and* $N = \sum_{j=1}^m N_j$.

The optimal solution and profits of the players in the decentralized scenario are denoted by the superscript SB. Unfortunately, similar to the scenario in previous studies (Kim et al. 2007), the customer's problem, as indicated by Eq. (5.8), is not generally quasi-convex because of the exponential function, and closed-form solutions cannot be obtained. The conditions that the solutions should satisfy are presented in Proposition 5.2.

Proposition 5.3 *An equilibrium exists for the decentralized game, in which all customers provide the same cost reimbursement to a given supplier, that is,* $\alpha_{ij}^{SB} = \alpha_{ik}^{SB}$ *for* $\forall j, \forall k = 1, 2, \ldots, m$.

The proof of Proposition 5.3 is presented in Appendix C.3. This proposition states that although customers may have different numbers of systems and different system profitability levels, they provide the same cost reimbursement to a particular supplier. This interesting and unexpected result contradicts the commonsense view stating that customers with a small number of systems or low profitability are free riders when suppliers establish a shared spare inventory for after-sales support.

This finding can be explained as follows. For convenience, we refer to customers with large scales, that is, a high value of $N_j P_j$, as large customers and customers with a lower value of $N_j P_j$ as small customers. In the context of PBC, for each customer, two terms represent the influence on supplier decisions (particularly decisions with respect to spare inventory) that directly affect system availability. One term is cost reimbursement, which exerts a pull effect on suppliers, and the other term is backorder penalties, which exert a push effect. In the Nash game among customers, a dominant strategy for large customers is selecting the cost reimbursement similar to that of small customers and requiring high backorder penalties to push the spare inventory to the desired level. As a result, all customers provide the same cost reimbursement.

Furthermore, according to the abovementioned analysis, the supplier receives less cost reimbursement for providing support services to different customers with different scales. This result is confirmed by the numerical results presented in Sect. 5.6. This finding encourages suppliers to establish dedicated inventories for their customers when the cost savings from the pooling effect is small. It is also consistent with the results of Chap. 4.

5.5 Supply Chain Coordination

In the PBC of Kim et al. (2007), the first-best solution can be achieved in the decentralized scenario *if and only if* $\alpha_i = 0, \upsilon_i = \theta^{FB}$, and $\omega_i = c_i s_i + \varepsilon_i + \theta^{FB'} E[B_i | s_i^{FB}] - 1/(2k_i)$ (for details, please see Proposition 3 on p. 1852 of Kim et al. 2007). Conversely, in our model with the fixed payment denoted as $c_i \mu_{ij}$, although the contract $\{\alpha_{ij} = 0, \rho_{ij} = 1\}$ can lead Eqs. (5.6) and (5.7) to the first-best solution, it is not the best choice for customers' profit according to Eqs. (5.9) and (5.11). In other words, we can assure that this solution does not lead the service supply chain to the maximal profit despite failure in obtaining the exact expressions of customers' decision. As a result, a coordination contract is needed by customers who want to induce suppliers to make decisions like those in the first-best scenario and to protect their own profit at the same time.

To achieve the perfect performance of the supply chain, customers provide some incentives to suppliers to gain more profits from the supply chain. A new contract can be adopted as follows:

$$T'_{ij} = d_{ij} N_j \left(c_i \mu_i + C'_i\right) / N + \alpha_{ij} \left(c_i s_i - a_i - C'_i\right) + \rho_{ij} P_j N_j \left(B'_i - B_i\right) / N,$$

where C'_i is a constant that represents the cost level of supplier i, and the customers provide a fixed payment to suppliers. Thereafter, a reconciling transfer payment is made according to the difference between customers' actual maintenance cost and the given level. B'_i is a given value of the amount of backorders, and it represents the basic level of the required availability of subsystem i. Customers provide a reward to suppliers to encourage them to maintain a high availability of the corresponding part. Finally, the coefficient d_{ij} is used to adjust the profits between customers and suppliers to ensure that they accept this contract and participate in the coordination.

With the new contract, the actions of supplier i tend to be the same as those presented in Proposition 5.1 because the new variables are constants and independent from the cost reduction effort and spare inventory. Therefore, their decisions are still shown as Eqs. (5.6) and (5.7).

The customers' decisions are changed after the suppliers' responses are made as shown as follows:

$$\alpha_{ij}^{CO} = \frac{\left(1 - \sum_{l=1}^{m} \alpha_{il}^{CO} - k_i \left(c_i s_i^{CO} - C_i'\right)\right) f_i \left(s_i^{CO}\right) \sum_{l=1}^{m} \rho_{il}^{CO} P_l N_l}{f_i \left(s_i^{CO}\right) \sum_{l=1}^{m} \rho_{il}^{CO} P_l N_l + N k_i c_i^2} + \frac{\left(1 - \rho_{ij}^{CO}\right) \left(1 - F_i \left(s_i^{CO}\right)\right) k_i c_i P_j N_j}{f_i \left(s_i^{CO}\right) \sum_{l=1}^{m} \rho_{il}^{CO} P_l N_l + N k_i c_i^2}, \tag{5.11}$$

$$\rho_{ij}^{CO} = 1 - \frac{N \alpha_{ij}^{CO} c_i \left(1 - F_i \left(s_i^{CO}\right)\right) + \left(B_{i0} - E[B_i | s_i^{CO}]\right) f_i \left(s_i^{CO}\right) \sum_{l=1}^{m} \rho_{il}^{CO} P_l N_l}{P_j N_j \left(1 - F_i \left(s_i^{CO}\right)\right)^2}, \tag{5.12}$$

where $s_i^{CO} = F_i^{-1}\left[\max\left\{1 - N\left(1 - \sum_{j=1}^{m} \alpha_{ij}^{CO}\right) c_i \Big/ \left(\sum_{j=1}^{m} \rho_{ij}^{CO} P_j N_j\right), F_i(0)\right\}\right]$.

In these equations, the superscript CO is used to indicate the optimal solutions and profits after the adoption of the coordination contract. In obtaining the terms of the coordination contract, the following conditions are held for any $i = 1, 2, \ldots, n$ and $j = 1, 2, \ldots, m$:

$$\alpha_{ij}^{CO} = 0, \tag{5.13}$$

$$\rho_{ij}^{CO} = 1, \tag{5.14}$$

$$\pi_i^{CO} \geq \pi_i^{SB}, \text{ and } \Pi_j^{CO} \geq \Pi_j^{SB}. \tag{5.15}$$

Considering Eqs. (5.13) and (5.14), we can find that $C_i' = c_i s_i^{CO} - 1/k_i$ and $B_i' = E[B_i | s_i^{CO}]$ are the solutions. Furthermore, according to Eq. (5.6), $s_i^{CO} = s_i^{FB}$ is now known. Thus, $C_i' = c_i s_i^{CO} - 1/k_i = C_i^{FB}$ and $B_i' = E[B_i | s_i^{FB}]$. Thereafter, the value of the coefficient d_{ij} can be obtained from the inequalities (5.15), and the participation of customers and suppliers in the coordination contract is ensured. The next section provides the necessary numerical studies due to the unprocurability of the closed form of the profits.

5.6 Numerical Illustrations

This section numerically illustrates the results and findings by employing the data used by Kim et al. (2007). For details, please see Sect. 4.3.2 of Chap. 4 as they are omitted here. In Sect. 5.6.1, we first show the results for an after-sales service supply chain with a single customer. Thereafter, we present the results with multiple

Table 5.1 Equilibrium for the supply chain with a single customer

Subsystem	$P = 50$					$P = 70$				
	a	b	c	d	e	a	b	c	d	e
α_i^{SB}	0.2798	1^-	0.1758	0.3606	0.3587	0.3195	1^-	0.2077	1^-	0.4076
ρ_i^{SB}	0.4625	0^+	0.6337	0.2257	0.3482	0.3614	0^+	0.4953	0^+	0.2702
a_i^{SB}	154.98	0^+	256.16	54.47	95.24	146.44	0^+	246.25	0^+	87.967
s_i^{SB}	9.0354	24.274	10.489	17.048	8.0874	9.8171	25.145	11.667	21.661	8.6848
A_i^{SB} (%)	98.369	99.317	97.326	98.573	98.926	98.682	99.379	97.901	99.409	99.120

Note 1^- indicates that the value approaches 1 from the left side, and 0^+ indicates that the value approaches 0 from the right side

customers in Sect. 5.6.2. Finally, we numerically show a coordination contract in Sect. 5.6.3.

5.6.1 Equilibrium with Single Customer

Based on the abovementioned parameters, the second best solutions for a single customer are presented in Table 5.1. We consider the two scenarios in which the profitabilities of the system are $P = 50$ and $P = 70$. The results indicate that the optimal inventory levels of the spare parts increase with the profitability of the system, and the system with high profitability thus performs at a high availability level. With high profitability, the customer does not set high backorder penalties for suppliers; instead, the customer specifies more cost compensation and less backorder penalties, which is an unexpected result. When profitability is high, the optimal availability of the system is also high. Consequently, backorder penalties carry less weight than cost compensation, and with the increase in profitability, the customer provides more cost compensation and less performance incentives to suppliers.

In both cases, the customer shares more maintenance costs and sets lower backorder penalties for suppliers with a low unit cost because the customer seeks to achieve higher availability from subsystems with a low unit cost which provide small marginal cost.

The optimal contract terms $\alpha_i^{SB} = 1^-$ and $\rho_i^{SB} = 0^+$ are of particular interest. According to Eq. (5.1) and Proposition 5.1, the optimal availability of a subsystem decreases with its unit cost and increases with the profitability of the assembled system. Accordingly, when designing contract terms, the customer provides greater cost reimbursement and less backorder penalties. Simultaneously, the provided cost reimbursement ratio decreases, and backorder penalties increase with the unit cost among the subsystems. In other words, for each subsystem, profitability $P^* > 0$ exists: when $P > P^*$, the cost compensation ratio of the system approaches 1 and backorder penalties approach 0 (e.g., see the values for subsystem b and subsystem d in Table 5.1). Consequently, the optimal spare inventory (which can also be considered as availability) of the subsystem approaches that of the centralized scenario

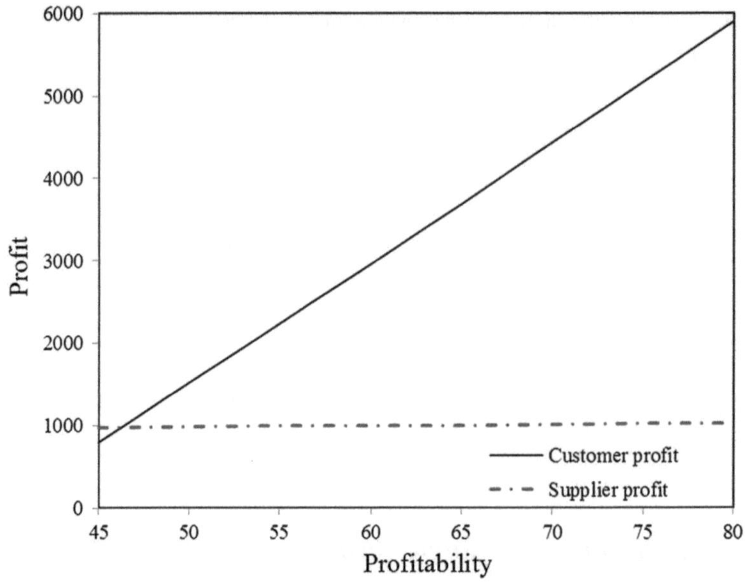

Fig. 5.2 Equilibrium profits of the players as a function of profitability

(Eq. (4.4) of Chap. 4), whereas the cost reduction effort of the corresponding supplier approaches 0.

With the optimal contract terms set by the customer and the inventory establishment and cost reduction effort investment provided by suppliers as presented and discussed above, the profits of the customer and suppliers in different demand seasons can be presented. For convenience, we further integrate the profits of the n suppliers into a single profit because all suppliers are independent of one another and do not affect the results.

Figure 5.2 presents the equilibrium profits of the customer and suppliers with varying profitabilities. This result illustrates that the influence of profitability on customer profit is significantly greater than its influence on supplier profit although both profits tend to increase as profitability increases. The customer increases the optimal availability requirement of the system to meet the demand for higher profit because the cost functions of suppliers are fixed and the marginal profit of the customer is more considerable when the profitability of the system is higher. The corresponding profit of the customer increases significantly with profitability growth. Consequently, in practice, the customer has strong incentives to perform market efforts to increase the profitability of its product and therefore its profit.

For suppliers, although profit increases with the profitability of the system, the increase is less significant than the increase in customer profit because the maintenance cost function structure of suppliers is not influenced by the profitability of the systems. In other words, the bullwhip effect caused by system profitability diminishes from the customer to suppliers. The results also indicate that the customer does

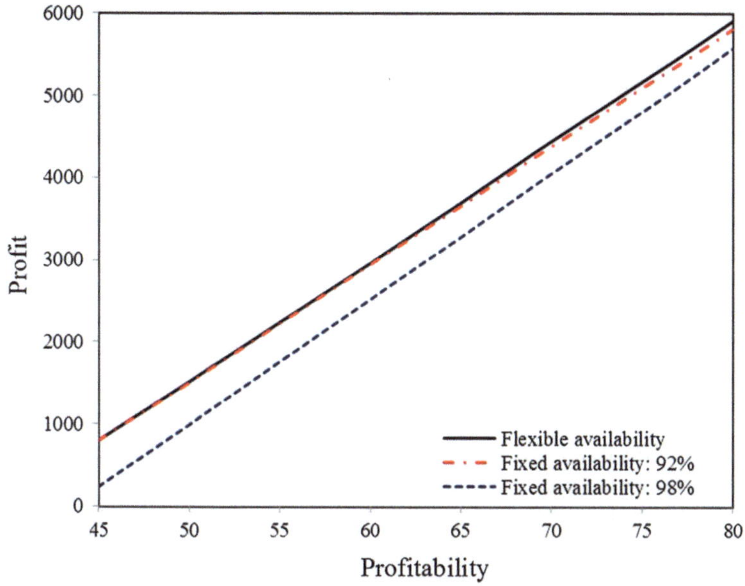

Fig. 5.3 Customer profit with different availability requirements

not obtain major profits from the supply chain when the profitability of the systems is low although the customer is the leader of the game.

Figure 5.3 illustrates the importance of the choice of target availability for the customer by comparison between the profit obtained with flexible availability (the optimal solution considering profitability) and that obtained with fixed availability (regardless of the profitability of the system). First, the results show that a fixed availability requirement is not optimal for differential profitability and is always inferior to the flexible policy, which considers profitability. Second, a higher fixed availability target is significantly more detrimental than a lower target. Figure 5.3 reveals that the profit loss from a fixed availability is significantly smaller when the availability requirement is fixed at 92 % than when the availability target is 98 %. This finding is similar to the conclusion in Chap. 4: the profit of the customer declines if a different (especially a higher) availability target is set regardless of the profitability of the system.

5.6.2 *Equilibrium with Multiple Customers*

We suppose that the 156 aircrafts are held by two customers. Customer 1 owns $N_1 = 100$ of the aircrafts with a corresponding profitability of $P_i = 47.2$. Customer 2 owns the remaining $N_2 = 56$ aircrafts with a profitability of $P_2 = 55$. Furthermore, for convenience, only two types of spare parts are relevant in this analysis: subsystem

Table 5.2 Equilibrium for the supply chain with multiple customers

	Shared inventory		Dedicated inventory	
	a	c	a	c
s^{SB}	8.5043	9.8730	–	–
s_1^{SB}	–	–	5.7681	6.4502
s_2^{SB}	–	–	3.5295	3.8768
α_1^{SB}	0.1751	0.1111	0.3507	0.2851
α_2^{SB}	0.1751	0.1111	0.4138	0.3763
ρ_1^{SB}	0.4914	0.6434	0.4616	0.5994
ρ_2^{SB}	0.2207	0.4536	0.4198	0.5170

a, which has a low unit cost of 21.52, and subsystem c, which has a high unit cost of 31.08. The results are presented in Table 5.2.

The results in Table 5.2 indicate that the supplier receives identical cost reimbursements from the two customers in the shared inventory scenario. Furthermore, the value of cost reimbursement is lower in the shared inventory scenario than in the dedicated inventory scenario. The results are consistent with the findings of Proposition 5.3 and indicate that a pooling effect exists in both cases in the decentralized scenario. However, the pooling effect for subsystem c is less than that for subsystem a because the unit cost of subsystem c is greater than that of subsystem a, whereas the average profitability of the systems is identical. In this example, the analysis further indicates that when the unit cost of a spare part for subsystem c exceeds 38, the pooling effect disappears. This result is also consistent with Proposition 4.6 in Chap. 4.

5.6.3 Supply Chain Coordination

The analysis in Sect. 5.5 indicates that the contract provided by the customer cannot induce suppliers to adopt the first-best solution in the decentralized scenario. In inducing suppliers to adopt the first-best solution and in achieving the full-channel coordination, a coordination contract is required. The optimal contract terms can be obtained according to Sect. 5.5.

Figure 5.4 shows the profit improvement from the coordination contract with only one customer and five suppliers. This figure indicates that when system profitability is low (smaller than 50), the improvement is attractive, and the coordination is worthwhile. Nevertheless, the profit improvement of the supply chain is not greater than 5 % in most cases (when system profitability is higher than 50) under the terms of the optimal coordination contract. In other words, the PBC performs at a high level in the decentralized scenario. The numerical results show that with the increase in the profitability of the system, the profit improvement achieved by the coordination

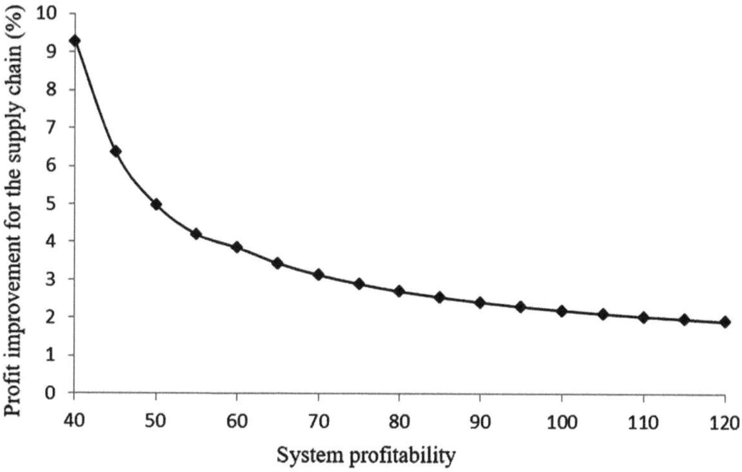

Fig. 5.4 Profit improvement of the coordination contract for the supply chain

contract decreases in the total supply chain. In other words, the contract in our model provides a high performance mechanism for the after-market service supply chains, and the coordination contract is not necessarily adopted.

5.7 Chapter Summary

This chapter investigates PBC in the after-sales service supply chains considering the profitability of the assembled system in the market and aims to determine the optimal contract terms for each subsystem that maximize the profit of customers. In addition, two practically important contract terms in the after-market service supply chains are analyzed: cost reimbursement and performance incentive. Through a multiple-leader-multiple-follower game, the optimal actions of customers and suppliers for the maintenance of the subsystems are analyzed. In addition, the profits of the players are surveyed, and a coordination contract is presented when it is desired by the supply chain in practice.

The main findings of this chapter are as follows. First, through the game model adopted in this study, the optimal required availability of the system is found to increase with the average profitability of the systems under an identical maintenance cost structure. Customers' profit decreases if they select a different (especially a higher) availability target for the suppliers regardless of the profitability of the system. Second, the analysis of multiple customers finds results similar to those in Chap. 4, in which the pooling effect does not always exist for a spare inventory setup. Moreover, the analysis also indicates that all customers provide the same cost reimbursement for suppliers when shared inventory is established even for customers with different numbers of systems and different profitabilities. However, the total cost reimbursement that the supplier receives in the shared inventory scenario is less than that in the dedicated inventory scenario. This result encourages suppliers to set dedicated inventories for their customers when the cost savings derived from the pooling effect is not significant. Third, additional findings are shown through the analysis of a single customer model (which can be considered in the dedicated inventory scenario). With the optimal contract terms, customers gain more profits because of the increasing profitability and the corresponding availability, and suppliers of the subsystems also benefit from the increasing availability requirement of the system. As a result, customers provide a high cost sharing ratio and a low backorder punishment to induce suppliers to set high inventory levels for the spare parts. However, the increasing cost sharing of customers reduces the level of cost reduction effort by suppliers because customers share the cost reduction effects but do not share the cost reduction investment. As a result, if the profitability of the system is high enough, then customers bear almost all of the maintenance costs for suppliers and provide a small punishment for the backorder spare parts, which infinitely approach zero. Furthermore, with a given profitability, customers of the system provide different contract terms to different suppliers according to the unit costs of the subsystems. The supplier with the lowest unit cost for a subsystem obtains the highest cost reimbursement and the lowest backorder punishment among all suppliers. This finding indicates that when the availability of the subsystems is at the same level, customers prefer to increase the availability of the subsystem with a low unit cost first to achieve the required availability of the system and then seek the maximal profit. Finally, the numerical studies in this study show that the performance-based contract performs at a high level, and the coordination contract is not always desired. In the examples in this study, the improvement of the coordination contract is less than 5 % of the total profit of the supply chain in the decentralized scenario in most cases. Therefore, the coordination effect is not the concern of the players because of some operational inconveniences, information-sharing risk, and other factors.

The analysis in this chapter is based on the following assumptions. The players are risk-neutral and share their information symmetrically. The repair lead time is constant, that is, the analysis of the maintenance service contract of the subsystems in the after-market supply chain is singly periodic. However, on the one hand, information asymmetry and risk aversion are common in operations management and have been given considerable attention in academic research (Chen 2005; Plambeck and Taylor 2006; Su and Zenios 2006; Lutze and Özer 2008). Hence, future studies can examine the problem with information asymmetry or risk problem to gain additional practical insights. On the other hand, the repair lead time of each subsystem is not constant in practice (Rappold and Van Roo 2009; Basten et al. 2012). That is, the performance contracts for suppliers of subsystems should be based on a whole-horizon decision process instead of a single-period problem. Therefore, future research can be conducted on contract design considering system profitability under stochastic repair lead time using dynamic decision methods.

References

Ahmadi-Khatir A, Fotuhi-Firuzabad M, Goel L (2009) Customer choice of reliability in spinning reserve procurement and cost allocation using well-being analysis. Electric Power Syst Res 79(10):1431–1440

Basten RJ, van der Heijden MC, Schutten J (2012) Joint optimization of level of repair analysis and spare parts stocks. Eur J Oper Res 222(3):474–483

Berkowitz D, Gupta JN, Simpson JT, McWilliams JB (2005) Defining and implementing performance based logistics in government. Def Acquis Rev J 11(3):255–267

Butler RW, Finelli GB (1993) The infeasibility of quantifying the reliability of life-critical real-time software. IEEE Trans Softw Eng 19(1):3–12

Cai KY, Wen CY, Zhang ML (1991) A critical review on software reliability modeling. Reliab Eng Syst Saf 32(3):357–371

Chen F (2005) Salesforce incentives, market information, and production/inventory planning. Manag Sci 51(1):60–75

Chung JY, Whang T (2011) The impact of low cost carriers on Korean island tourism. J Transp Geogr 19(6):1335–1340

Cohen MA, Agrawal N, Agrawal V (2006a) Achieving breakthrough service delivery through dynamic asset deployment strategies. Interfaces 36(3):259–271

Cohen MA, Agrawal N, Agrawal V (2006b) Winning in the aftermarket. Harv Bus Rev 84(5):129

Dai YS, Xie M, Poh KL (2008) Availability modeling and cost optimization for the grid resource management system. IEEE Trans Syst, Man Cybern, Part A: Syst Hum 38(1):170–179

Dennis MJ, Kambil A (2003) Service management: building profits after the sale. Supply Chain Manag Rev 7(3):42–48

Gaiardelli P, Saccani N, Songini L (2007) Performance measurement of the after-sales service network evidence from the automotive industry. Comput Ind 58(7):698–708

Gruneberg S, Hughes W, Ancell D (2007) Risk under performance-based contracting in the UK construction sector. Constr Manag Econ 25(7):691–699

Guajardo JA, Cohen MA, Kim SH, Netessine S (2012) Impact of performance-based contracting on product reliability: an empirical analysis. Manag Sci 58(5):961–979

Gustafsson A, Brax S, Witell L, Ng IC, Nudurupati SS (2010) Outcome-based service contracts in the defence industry-mitigating the challenges. J Serv Manag 21(5):656–674

Jin B, Sternquist B (2003) The influence of retail environment on price perceptions: an exploratory study of US and Korean students. Int Mark Rev 20(6):643–660

Kang WH, Song J, Gardoni P (2008) Matrix-based system reliability method and applications to bridge networks. Reliab Eng Syst Saf 93(11):1584–1593

Kim SH (2011) Strategic reliability investments in multi-indenture supply chains. Technical report, Working paper, Yale University

Kim SH, Cohen MA, Netessine S (2007) Performance contracting in after-sales service supply chains. Manag Sci 53(12):1843–1858

Kim SH, Cohen MA, Netessine S, Veeraraghavan S (2010) Contracting for infrequent restoration and recovery of mission-critical systems. Manag Sci 56(9):1551–1567

Kim SH, Cohen MA, Netessine S (2011) Reliability or inventory? Analysis of product support contracts in the defense industry

Lutze H, Özer Ö (2008) Promised lead-time contracts under asymmetric information. Oper Res 56(4):898–915

Mettas A (2000) Reliability allocation and optimization for complex systems. Proceedings of the annual reliability and maintainability symposium, 2000. IEEE, pp 216–221

Mirzahosseinian H, Piplani R (2011) A study of repairable parts inventory system operating under performance-based contract. Eur J Oper Res 214(2):256–261

Ng IC, Maull R, Yip N (2009) Outcome-based contracts as a driver for systems thinking and service-dominant logic in service science: evidence from the defence industry. Eur Manag J 27(6):377–387

Oikonomou KN (1993) System reliability calculations based on incomplete information. IEEE Trans Syst, Man Cybern 23(5):1235–1254

Oliva R, Kallenberg R (2003) Managing the transition from products to services. Int J Serv Ind Manag 14(2):160–172

Oppermann M, Cooper M (1999) Outbound travel and quality of life: the effect of airline price wars. J Bus Res 44(3):179–188

Plambeck EL, Taylor TA (2006) Partnership in a dynamic production system with unobservable actions and noncontractible output. Manag Sci 52(10):1509–1527

Rappold JA, Van Roo BD (2009) Designing multi-echelon service parts networks with finite repair capacity. Eur J Oper Res 199(3):781–792

Robidoux R, Xu H, Xing L, Zhou M (2010) Automated modeling of dynamic reliability block diagrams using colored petri nets. IEEE Trans Syst, Man Cybern, Part A: Syst Hum 40(2):337–351

Roels G, Karmarkar US, Carr S (2010) Contracting for collaborative services. Manag Sci 56(5):849–863

Şen A (2008) The US fashion industry: a supply chain review. Int J Prod Econ 114(2):571–593

Su X, Zenios SA (2006) Recipient choice can address the efficiency-equity trade-off in kidney transplantation: a mechanism design model. Manag Sci 52(11):1647–1660

Wang Y, Li L (2012) Effects of uncertainty in both component reliability and load demand on multistate system reliability. IEEE Trans Syst, Man Cybern, Part A: Syst Hum 42(4):958–969

Xing L, Levitin G (2011) Combinatorial algorithm for reliability analysis of multistate systems with propagated failures and failure isolation effect. IEEE Trans Syst, Man Cybern, Part A: Syst Hum 41(6):1156–1165

Yamada S, Osaki S (1985) Cost-reliability optimal release policies for software systems. IEEE Trans Reliab 5:422–424

Zio E (2009) Reliability engineering: old problems and new challenges. Reliab Eng Syst Saf 94(2):125–141

Chapter 6
Conclusions

Abstract This chapter presents the summary of the book, the limitations of the research, and future research directions.

6.1 Book Summary

Along with the development of the Internet and electronic commerce, the logistics and supply chain have become one of the most popular research topics since the end of the last century. In this topic, warehousing systems as important nodes in supply chains receive great attention from scholars. In practice, the rapid development of B2C and C2C e-commerce brings new challenges to warehouses in order response. In this business environment, this book examines the most important issues related to the order response efficiency of warehousing systems: the storage policies of the system and the availability of the S/R machine. According to these two research issues, the research questions and contributions of this book are summarized as follows.

First, as an effective strategy, the class-based storage policy is widely studied in the literature and applied in practice. Existing studies have provided many significant suggestions for warehousing operations. One of the most famous results shows that the full turnover-based policy is the best solution for a warehousing system to achieve the minimal expected one-way travel time, and warehouse managers are advised to divide items into different items according to their turnovers in as many classifications as possible. However, this suggestion is inconsistent with practice as usually only a few (three to five) classes are implemented. This book revisits the existing literature and finds that all the studies have been conducted with a basic assumption: "the required storage space of all the items equals their average inventory level". This assumption is valid only if an infinite number of items are stored in each storage region. It is incompatible with reality in which the full turnover-based storage policy is conducted because only one item exists in each class in this situation. The incompatibility between the assumption and the storage policy ignores the influence of RSS on travel time or distance. To investigate the influence of RSS on the item classification of the class-based storage policy, Part I of this book revisits the travel time or distance model by relaxing the assumption that "the required storage space of

X. Guo, *Storage Policies and Maintenance Support Strategies in Warehousing Systems*, Springer Theses, DOI 10.1007/978-981-10-1448-2_6

all the items equals their average inventory level" and by considering a finite number of items in a warehousing system. The contributions of this part are summarized as follows:

(1) The influence of the infinity of stored items in a warehousing system is considered when modeling the travel time. The results explain the non-adoption of the famous full turnover-based storage policy in practice. Conventional research only considers the improvement of the classification of items according to their turnovers but ignores the expansion of RSS when the number of classes increases. The results of this book indicate that, normally, a small number of classes are optimal for warehousing systems to minimize the average one-way travel time.
(2) Travel time is insensitive to the number of classes in a wide range beyond the optimum. For example, in the numerical study in Chap. 2, three to eight classes are near optimal, and travel time is not sensitive to the number of classes. This result suggests that warehouse managers can freely change the number of classes when necessary (e.g., for space use purposes).
(3) This book also studies the influence of demand structure (the skewness of ABC demand curves) on the performance of storage policies by considering the realistic RSS measured in a number of locations. Conventional research has indicated that the average one-way travel time under the random storage policy is the same for different ABC curves. Considering the influence of RSS, we find significant gaps exist between travel times for the same storage policy with different ABC curves. For example, for the random storage policy, an approximately 40 % gap exists between travel times based on the 20 %/20 % and 20 %/90 % curves. This gap is a result of the differences in RSSs with different ABC curves.
(4) Based on the theoretical results, this book provides the optimal classification strategy (including the number of classes and corresponding class boundaries) for AS/RSs and traditional unit-load warehouses with aisles. In addition, the optimal warehouse shape factors for different ABC demand curves are provided.

Second, the response efficiency of warehousing systems depends not only on the storage policies but also on the availability of the S/R machines, thus making maintenance support important for critical equipment (or systems). In the environment where firms transform from manufacturing businesses to service businesses, support and maintenance service in the after-sales market has become an important component of economic development, and this service usually provides a higher profit rate to suppliers than the sales of the original products. Minimizing the maintenance cost with a given availability requirement is considered as the major objective in the topic of after-sales service supply chain management. According to the existing literature, the performance-based contract has been paid considerable attention by customers because it can realize the minimum maintenance cost with a given availability target of the systems. However, in practice, the major objective of an enterprise is not to minimize its operational or maintenance cost but to maximize its profit. Therefore, Part II of this book introduces a new contract considering the profitability (which has a strong correlation with the availability requirement of the equipment) of the equipment. The contributions of this part are summarized as follows:

(1) A maintenance contract considering the profitability of equipment is proposed, and the corresponding optimal decisions are derived. The results show that the optimal required availability of the equipment is found to increase with its profitability in the market. Decisions based on a given availability target may reduce the profit of the enterprise, especially when the enterprise seeks high availability regardless of the profitability of the system.
(2) A multi-leader-multi-follower sequential game is analyzed to study the interactions of customers and suppliers in after-sales service supply chains. From the perspective of customers, all of them provide the same cost reimbursement for supplier when a shared inventory is established even for customers with different numbers of systems and different profitabilities. This finding is derived because in the Nash game among customers, selecting the same cost reimbursement as that of the small customers and requiring high backorder penalties to push the spare inventory to the desired level are dominant strategies for large customers with more systems or higher profitabilities.
(3) From the perspective of service providers, establishing a shared spare inventory for all customers is not always better than setting dedicated inventories for each customer. A threshold exists: when the unit cost of a spare part is lower than the threshold, a shared inventory saves cost for the supplier, whereas when the unit cost of a spare part is higher than the threshold, suppliers should set a dedicated spare inventory for each customer.

6.2 Directions for Future Research

This book examines the storage policies and maintenance support strategies for warehousing systems by considering the finiteness of the number of items stored in a warehouse and the profitability of the equipment. The current research is limited, but necessary assumptions and future research can be performed in the following directions:

(1) The research on class-based storage with a finite number of items is conducted in AS/RS based on a single-command mode. Further research based on an operational mode with dual-command can also deliver interesting insights. Other warehousing systems, such as order-picking systems, are also considerable.
(2) The research on the performance of class-based storage in unit-load warehouses is conducted on the basis of a traditional layout with parallel storage aisles. Similar research can be extended to warehouses with flying-V or fishbone layouts. Furthermore, the current research only considered the across-aisle boundary between classes. Within-aisle boundary and diagonal boundary are also worth studying to deduce more managerial insights for practice.

(3) Many considerable directions for future research also exist on the topic of maintenance support. For example, the current research assumes that the players are risk-neutral and have symmetric information. Information asymmetry and risk aversion are common in operations management and have been given considerable attention in academic research. Therefore, future studies can work on the problem with information asymmetry or risk problem to gain additional practical insights.
(4) The repair lead time of each subsystem is considered as a constant in this research, while it may vary in practice. Thus, the performance contracts for the suppliers of subsystems should be based on a whole-horizon decision process instead of a single-period problem. Hence, future research on contract design considering system profitability under stochastic repair lead time through dynamic decision methods is worth conducting.

Erratum to: Storage Policies and Maintenance Support Strategies in Warehousing Systems

Xiaolong Guo

Erratum to:
X. Guo, *Storage Policies and Maintenance Support Strategies in Warehousing Systems*, Springer Theses, DOI 10.1007/978-981-10-1448-2

The book was inadvertently published with an incorrect affiliation. The correct affiliation is, School of Management, University of Science and Technology of China, Hefei, People's Republic of China. The book and the frontmatter has been updated.

The updated original online version for this book can be found at
DOI 10.1007/978-981-10-1448-2

X. Guo (✉)
School of Management, University of Science and Technology of China, Hefei,
People's Republic of China
e-mail: gxl@ustc.edu.cn

X. Guo, *Storage Policies and Maintenance Support Strategies in Warehousing Systems*, Springer Theses, DOI 10.1007/978-981-10-1448-2_7

Appendix A
Finding ε Using Simulation

This appendix consists of two parts. It first gives the method to obtain a single ε value for a given item i_0 ($0 < i_0 < 1$) and s by simulation (i_0 represents the index of the item with the highest turnover among the N_k items allocated in class k). Then it tests the impact of s and i_0 on ε. The simulation details are based on stochastic demand. The same process can be applied to the deterministic case by setting $\sigma_i = 0$.

A.1 Obtaining ε for a Given i_0 and s by Simulation

According to the parameters given in Sect. 2.5.2.2, set N, $\lambda(i)$ ($D(i)$ in the deterministic demand situation), l_i, α_i, and we can get ss_i, r_i, S_i, and $Q(i)$ accordingly. Since the items are normalized in the Model BM, we also do this in Model DSM by setting $i_0 \in \{1/N, 2/N, \ldots, 1\}$. For the given class k of i_0, the number of items in this class is $N_k \in \{1, 2, \ldots, (1 - i_0)N + 1\}$. We use the following steps to obtain the value of ε.

(1) Generate a required storage space, $\hat{a}_i(N_k)$, at a given $N_k \in \{1, 2, \ldots, (1 - i_0)N + 1\}$. This step consists of four substeps.

   (i) Initialize inventory level for each item $i \in \{i_0, i_0 + 1/N, \ldots, (N_k - 1)/N\}$; the initial inventory level of each item is generated as a random number (integer for Model DSM) between reorder point r_i and order-up-to level S_i.

   (ii) Obtain the inventory level of item $i \in \{i_0, i_0 + 1/N, \ldots, (N_k - 1)/N\}$, as a function of time by discrete event simulation of demand and replenishments according to the (r_i, S_i) replenishment policy over a period of 100 time units (years). The simulated time period represents an average of 10,000 units of demand and 500 replenishments per item.

   (iii) Obtain the total required storage space, $\hat{a}_i(N_k)$, for class k. Add all the inventory levels of the N_k items in the class to obtain the total inventory level of all N_k items as a function of time. $\hat{a}_i(N_k)$ equals the maximum of the total inventory level over time.

X. Guo, *Storage Policies and Maintenance Support Strategies in Warehousing Systems*, Springer Theses, DOI 10.1007/978-981-10-1448-2

(iv) The required storage space for item $i \in \{i_0, i_0 + 1/N, \ldots, (N_k - 1)/N\}$ in the class can be obtained by dividing the total space over the items,

$$\hat{a}_i(N_k) = \left(\hat{a}(N_k) - \sum_{l=i_0}^{i_0+(N_k-1)/N} ss_l\right) \frac{S_i - r_i}{\sum_{l=i_0}^{i_0+(N_k-1)/N}(S_i - r_i)} + ss_i,$$

where $(S_i - r_i) \Big/ \sum_{l=i_0}^{i_0+left.(N_k-1)right/N}(S_i - r_i)$ is the weighted value of item i among all N_k items for the storage space excluding the safety stock.

(2) Repeat step 1 and generate more instances of a_i, for the given N_k. 100 instances are generated in our simulations. The average value of the 100 instances will be used for estimating ε.
(3) Repeat steps 1 and 2 for every $N_k \in \{1, 2, \ldots, (1 - i_0)N + 1\}$. Figure A.1 shows the simulated results based on the parameters given in Sect. 2.5.2.2, for $s = 0.222$ normalized by $(S_i - r_i)$.
(4) With the sample data of $\hat{a}_i(N_k)$ generated with different values of N_k, we can estimate ε by using the least squares method.

(i) Calculate the sum of the squared errors (SSE) between the data values and the values derived from the function in Eq. (2.30) (Eq. (2.10) for deterministic demand) given in the following equation:

$$SSE = \sum_{N_k=1}^{M} \left(\hat{a}_i(N_k) - a_i(N_k)\right) 2, \tag{A.1}$$

where $\hat{a}_i(N_k)$ is the simulated data of a_i at N_k (e.g. a data point in Fig. A.1), and $a_i(N_k)$ is the estimate of a_i by using the Eq. (2.30) (Eq. (2.10) for deterministic demand).

(ii) By setting the first-order derivative of SSE with respect to ε equals zero, $\partial SSE/\partial\varepsilon = 0$, we obtain the best value of ε.

Using this method, in the next part, ε appears to be between 0.17 and 0.24 for most cases.

A.2 Variation of ε

Equations (2.10), (2.30), and (A.1) show that the value of ε is not influenced by the safety stock. Hence, the service level does not influence the value of ε. Although ε may be influenced by factors like s and i_0, we surprisingly found ε is little influenced by these factors. Figure A.2 shows that ε is mostly between 0.17 and 0.24 when $i_0 < 0.8$ ($i_0 > 0.8$ cannot provide enough sample data for the simulation, therefore we only use the result with $i_0 < 0.8$) with the parameters given in Sect. 2.5.2.2,

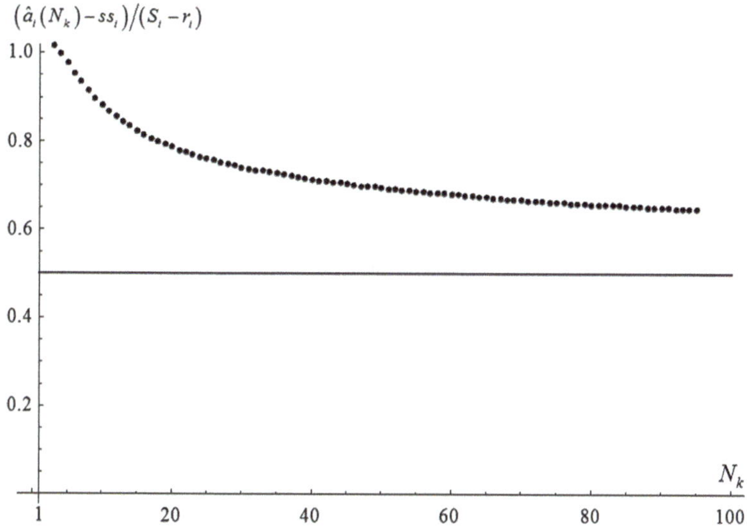

Fig. A.1 Sample data set for calculating ε

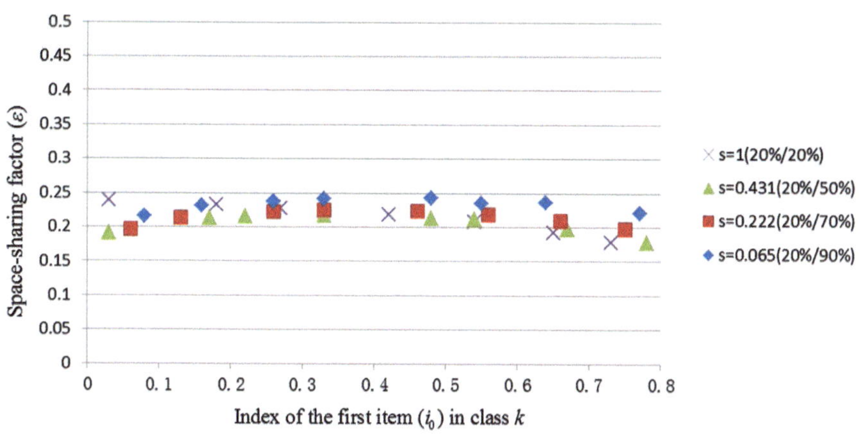

Fig. A.2 ε values for different simulated cases, with varying s and i_0

and does not depend heavily on s and i_0. We also have simulated different demand uncertainties. The results show that demand variability will reduce the space sharing efficiency, but fortunately, the affect is small and we can omit it here. Results for Bender's ABC-demand curve show the value of ε is again almost between 0.15 and 0.25. We omit the figures here since they are similar to Fig. A.2. Accordingly, we choose the average value of $\varepsilon = 0.22$ in the numerical study section.

Appendix B
Details of Solution Methodology in Sect. 2.3.3

To solve Model SM, we reformulate it as a dynamic programming model, and define it as follows. (i) Define k as the number of the stage; it corresponds to the kth storage class. (ii) Define Y_k as the state at stage k; it is defined by the total number of items in classes $1, 2, \ldots, k$, $Y_k \in [k, N - (n - k)]$. Y_k is lower bounded by k since at least one item is needed for each class in the previous classes $1, 2, \ldots, k$. Y_k is not larger than $N - (n - k)$ since at least $(n - k)$ items are stored in classes $k + 1, k + 2, \ldots, n$. (iii) Define N_k as the decision variable at stage k, it is the number of items in class k and its corresponding decision domain is $N_k \in [1, Y_k - k + 1]$. Note that $N_k \leq Y_k - k + 1$ because Y_k items are contained in classes up to k, and at least $(k - 1)$ items are needed to construct the previous $(k - 1)$ classes.

With the definitions given so far, $Y_k = N_k + Y_{k-1}$ is the state transfer function between the two states Y_{k-1} and Y_k, and $f_k(Y_k)$ is the evaluation function at stage k with given Y_k; it represents the average one-way travel time for Y_k items in the first k classes. $f_k(Y_n = N)$ equals T_n.

Then the recursive function for $f_k(Y_k)$ can now be written as follows:

$$f_k(Y_k) = f_{k-1}(Y_{k-1}) + \frac{2\left(R_k^3 - R_{k-1}^3\right)}{3\left(R_k^2 - R_{k-1}^2\right)}\left(\left(\frac{Y_k}{N}\right)^s - \left(\frac{Y_{k-1}}{N}\right)^s\right), \tag{B.1}$$

where $f_0(0) = 0$, $Y_0 = 0$, and $N_0 = 0$.

The minimum one-way travel time at stage $k > 1$ based on a known optimal solution at stage $(k - 1)$ can be found via solving the following:

$$f_k^*(Y_k) = \min_{1 \leq N_k \leq Y_k - k + 1}\left\{f_{k-1}^*(Y_{k-1}) + \frac{2\left(R_k^3 - R_{k-1}^3\right)}{3\left(R_k^2 - R_{k-1}^2\right)}\left(\left(\frac{Y_k}{N}\right)^s - \left(\frac{Y_{k-1}}{N}\right)^s\right)\right\}, \tag{B.2}$$

where R_k is determined by Eq. (2.19) and $Y_{k-1} = Y_k - N_k$.

$T_n = f_n^*(Y_n = N)$ gives the minimum travel time by optimizing the boundaries of n classes. To find T_n, we can iteratively solve $f_k^*(Y_k)$ for $k = 1, 2, \ldots, n$, respectively

X. Guo, *Storage Policies and Maintenance Support Strategies in Warehousing Systems*, Springer Theses, DOI 10.1007/978-981-10-1448-2

with initial conditions $f_0(0) = 0$ and $Y_0 = 0$. More specifically, to obtain $f_k^*(Y_k)$, we can numerically calculate its value via Eq. (B.2) for $Y_k = k, k+1, \ldots, N-n+k$ once $f_{k-1}^*(Y_{k-1})$ for all known $Y_{k-1} = k-1, k, \ldots, N-n+k-1$. Therefore, as $f_0(0) = 0$, $Y_0 = 0$, and $N_0 = 0$ are known, we can obtain $f_k^*(Y_k)$ for $k = 1, 2, \ldots, n$ iteratively to obtain $f_n^*(Y_n = N)$. To find N_k^* at stage k, corresponding to $f_k^*(Y_k)$, we need to enumerate N_k from 1 to $Y_k - k + 1$.

The computational complexity can now be determined. Because $k = 1, 2, \ldots, n$, $Y_k = k, k+1, \ldots, N-n+k$, and $N_k = 1, 2, \ldots, Y_k - k + 1$ are enumerated, respectively, the computational complexity of this algorithm to find $f_n^*(Y_n = N)$ is $O(N^2)$ and it is repeated n times.

The optimal number of classes, n^*, can be obtained by $n^* = \arg\min_{1 \le n \le N} \{f_n^*(N)\}$. Note that to optimize n^*, we do not need to successively calculate all of $f_1^*(N)$, $f_2^*(N), \ldots, f_N^*(N)$. In fact, $f_n^*(N)$ for any $n \le N$ can be simultaneously obtained when we compute $f_N^*(N)$ and additionally allow $Y_k = k, k+1, \ldots, N$ in Eq. (B.2). Allowing Y_k to equal N, we can obtain $f_k^*(N)$ (that is, the optimal travel time for a system with k classes) at stage k in the process of calculating $f_N^*(N)$. Therefore, the computational complexity for finding n^* is $O(N^3)$, which is the same as computing $f_N^*(N)$. Therefore, for realistically sized storage racks, the optimal solution can be found very fast.

Appendix C
Proofs

This chapter presents the proofs required in this book. Appendix C.1 gives the proof for Theorem 3.1 in Chap. 3; Appendix C.2 gives the proof of Proposition 4.6 in Chap. 4; and Appendix C.3 delivers the proof of Proposition 5.3 in Chap. 5.

C.1 Proof of Theorem 3.1

Considering that $y_k = L_k \big/ (4x+2)$ and $M_n = \sum_{k=1}^{n} \left(\frac{L_k + L_{k-1}}{2} \left(\left(\frac{i_k}{N}\right)^s - \left(\frac{i_{k-1}}{N}\right)^s \right) \right)$, the objective function of T_n can be rewritten as

$$
\begin{aligned}
T_n &= \frac{x(x+1)w}{2x+1} + \sum_{k=1}^{n} \left(\frac{y_k + y_{k-1}}{2} \left(\left(\frac{i_k}{N}\right)^s - \left(\frac{i_{k-1}}{N}\right)^s \right) \right) \\
&= \frac{x(x+1)w}{2x+1} + \frac{1}{4x+2} \sum_{k=1}^{n} \left(\frac{L_k + L_{k-1}}{2} \left(\left(\frac{i_k}{N}\right)^s - \left(\frac{i_{k-1}}{N}\right)^s \right) \right) \\
&= \frac{x(x+1)w}{2x+1} + \frac{M_n}{4x+2} \\
&= \frac{x(x+1)}{4} + \frac{2M_n - w}{4(2x+1)}.
\end{aligned}
$$

Since $x \geq 0$ and $2M_n > w$ in practice, according to $\mu^2 + v^2 \geq 2\mu v,\ \forall \mu, v$, we have $T_n \geq \sqrt{w(2M_n - w)} \big/ 2$, and $T_n = \sqrt{w(2M_n - w)} \big/ 2$ if and only if $x = \sqrt{(2M_n - w) \big/ (4w)} - 0.5$.

Next, we give the proof of $M_n < \sqrt{2NKA}$ and $x^* < \sqrt{\sqrt{2NKA} \big/ (2w)}$.

X. Guo, *Storage Policies and Maintenance Support Strategies in Warehousing Systems*, Springer Theses, DOI 10.1007/978-981-10-1448-2

$$\begin{aligned} M_n &= \sum_{k=1}^{n}\left(\frac{L_k+L_{k-1}}{2}\left(\left(\frac{i_k}{N}\right)^s-\left(\frac{i_{k-1}}{N}\right)^s\right)\right)\\ &< \sum_{k=1}^{n}\left(L_k\left(\left(\frac{i_k}{N}\right)^s-\left(\frac{i_{k-1}}{N}\right)^s\right)\right)\\ &< \sum_{k=1}^{n}\left(L_n\left(\left(\frac{i_k}{N}\right)^s-\left(\frac{i_{k-1}}{N}\right)^s\right)\right)\\ &< L_n. \end{aligned}$$

According to Eq. (3.8), $M_n < L_n \le L_N$, where L_N is the RSS for the warehouse with N classes each with only one item in the class and,

$$L_N = \sqrt{2KA}\sum_{i=1}^{N}\sqrt{(i/N)^s - ((i-1)/N)^s}.$$

Since $G(i) = (i/N)^s$ is a concave function of i, and $\sum_{i=1}^{N}((i/N)^s - ((i-1)/N)^s) = 1$, we find, $L_N \le \sqrt{2KA}\sum_{i=1}^{N}\sqrt{i/N - (i-1)/N} = \sqrt{2NKA}$. That is to say, the maximum RSS is achieved only if the warehouse stores N items with identical demand (a 20%/20% ABC-demand curve). Therefore, $M_n < L_n \le L_N \le \sqrt{2NKA}$. Using $x^* = \sqrt{(2M_n - w)/(4w)} - 0.5$, we find $x^* < \sqrt{\sqrt{2NKA}/(2w)}$.

Finally, since L_n is the total RSS for an n-class-based storage warehouse with $2x+1$ storage aisles, the required storage sections in depth can be obtained as $L_n/(4x+2)$. Consequently, the optimal warehouse shape factor (the ratio of warehouse width to depth) can be obtained as $r_n^* = w(2x^*+1)/(L_n/(4x^*+2))$. Considering $x^* = \sqrt{(2M_n - w)/(4w)} - 0.5$, it follows $r_n^* = 2(2M_n - w)/L_n$. Furthermore, recall $M_n < L_n$ for any number of classes under any ABC-demand curve, there is $r_n^* < 4$.

C.2 Proof of Proposition 4.6

Take subsystem i as an example. For symmetric enterprises, all the m enterprises hold the same number of systems, N_j, and have the same profitability, P_j. According to Propositions 4.1 and 4.4, the optimal dedicated inventory levels for each enterprise is $s_{ij}^d = F_{ij}^{-1}\left[\max\{1 - c_i/P_j, F_{ij}(0)\}\right]$, where $F_{ij}(\cdot)$ is the cdf of the demand of enterprise j for component i with $O_{ij} = O_i N_j/N$. Thus, the total inventory level for component i in this case is $s_i^d = m s_{ij}^d$. While in the situation with alliance, the total inventory level for component i is $s_i^{PI} = F_i^{-1}\left[\max\{1 - c_i/P_j, F_i(0)\}\right]$.

Table C.1 Monotonicity of $g(x)$

Interval	$-\infty$	$(-\infty, x_1)$	(x_1, μ_{ij})	μ_{ij}	(μ_{ij}, x_2)	$(x_2, +\infty)$	$+\infty$
Monotonicity or value of $g(x)$	>0	Increasing >0	Decreasing >0	0	Decreasing <0	Increasing <0	<0

To investigate the pooling effect of the alliance, we first define $g(x) = F_{ij}(x) - F_i(mx)$, where $F_{ij}(\cdot)$ is the cdf of the demand of each single enterprise with mean value μ_{ij}, and $F_i(\cdot)$ is the cdf of the demand of the alliance with mean value $\mu_i = m\mu_{ij}$.

The first-order derivative of $g(x)$ is

$$g'(x) = \frac{1}{\sqrt{2\pi\mu_{ij}}}\exp\left\{-\frac{(x-\mu_{ij})^2}{\mu_{ij}}\right\} - \frac{m}{\sqrt{2\pi m\mu_{ij}}}\exp\left\{-\frac{(mx-m\mu_{ij})^2}{m\mu_{ij}}\right\}.$$

Then, from $g'(x) = 0$, $x_1 = \mu_{ij} - \sqrt{\frac{\mu_{ij}\ln\sqrt{m}}{m-1}}$ and $x_2 = \mu_{ij} + \sqrt{\frac{\mu_{ij}\ln\sqrt{m}}{m-1}}$ are realized.

Hence, we have $g'(x) > 0$ if $x \in (-\infty, x_1) \cup (x_2, +\infty)$ and $g'(x) < 0$ if $x \in (x_1, x_2)$.

Furthermore, we can also proof that $g(x)|_{x\to-\infty} > 0$ and $g(x)|_{x\to+\infty} < 0$ as follows. From the definition of cumulative distribution function and probability density function, we have $g(x)|_{x\to-\infty} = f_{ij}(x) - f_i(mx)$. While $f_{ij}(x) - f_i(mx) > 0$ exists for any $x \in R$. Hence, $g(-\infty) > 0$ exists. $g(+\infty) < 0$ can be obtained similarly. As a result, $g(x) > 0$ if $x < \mu_{ij}$ and $g(x) < 0$ if $x > \mu_{ij}$.

Therefore, we can get the monotonicity of $g(x)$ as shown in Table C.1, which tells that $g(x) > 0$ when $x < \mu_{ij}$ and $g(x) < 0$ when $x > \mu_{ij}$.

According to the optimal solution presented in Propositions 4.1 and 4.4, we know that $s_{ij}^d < \mu_{ij}$ if $c_i > P_j/2$, and $s_{ij}^d > \mu_{ij}$ if $c_i < P_j/2$. Recall that $s_i^{PI} = F_i^{-1}\left[\max\{1 - c_i/P_j, F_i(0)\}\right]$ and $s_{ij}^d = F_{ij}^{-1}\left[\max\{1 - c_i/P_j, F_{ij}(0)\}\right]$, that is, $F_i[s_i^{PI}] = F_{ij}[s_{ij}^d] = 1 - c_i/P_j$ in the scenario with symmetric enterprises. Furthermore, both situations that the spare part inventory is either shared or dedicated are providing the same availability of the systems. Consequently, we can see that the pooling effect exists only when $c_i < P_j/2$.

C.3 Proof of Proposition 5.3

According to the analysis of Sect. 5.4, the optimal solutions follow the constraint conditions that are shown as Eqs. (5.6), (5.9), and (5.10). We first introduce the solution methodology of solving the equation set formed by these equations.

Denote $N = \sum_{j=1}^{m} N_j$, $\boldsymbol{\alpha}_i = \sum_{j=1}^{m} \alpha_{ij}$, and $\boldsymbol{\rho}_i = \sum_{j=1}^{m} \rho_{ij} P_j N_j$. Then, on one hand, from Eq. (5.9), we have

$$\begin{aligned}\boldsymbol{\alpha}_i &= \frac{m(1-\boldsymbol{\alpha}_i - k_i c_i s_i) f_i(s_i)\boldsymbol{\rho}_i + (1-F_i(s_i))\, k_i c_i \sum_{j=1}^{m} P_j N_j - (1-F_i(s_i))\, k_i c_i \boldsymbol{\rho}_i}{f_i(s_i)\boldsymbol{\rho}_i + N k_i c_i^2} \\ &= \frac{m(1-k_i c_i s_i) f_i(s_i)\boldsymbol{\rho}_i + (1-F_i(s_i))\, k_i c_i \sum_{j=1}^{m} P_j N_j - (1-F_i(s_i))\, k_i c_i \boldsymbol{\rho}_i}{(m+1) f_i(s_i)\boldsymbol{\rho}_i + N k_i c_i^2}.\end{aligned} \tag{C.1}$$

On the other hand, from Eq. (5.14), there is

$$\boldsymbol{\rho}_i = \frac{(1-F_i(s_i))^2 \sum_{j=1}^{m} P_j N_j - N\boldsymbol{\alpha}_i c_i (1-F_i(s_i))}{(1-F_i(s_i))^2 + m\,(B_{i0} - E\,[B_i|s_i])\, f_i(s_i)}. \tag{C.2}$$

Furthermore, from Eq. (5.6), we have,

$$\boldsymbol{\rho}_i = \frac{N c_i (1-\boldsymbol{\alpha}_i)}{1-F_i(s_i)}. \tag{C.3}$$

By substituting Eq. (C.3) into (C.2), we can further rewrite $\boldsymbol{\alpha}_i$ and $\boldsymbol{\rho}_i$ as

$$\boldsymbol{\alpha}_i = 1 + \frac{N c_i (1-F_i(s_i))^2 - (1-F_i(s_i))^3 \sum_{j=1}^{m} P_j N_j}{m N c_i\, (B_{i0} - E\,[B_i|s_i])\, f_i(s_i)}. \tag{C.4}$$

$$\boldsymbol{\rho}_i = \frac{(1-F_i(s_i))^2 \sum_{j=1}^{m} P_j N_j - N c_i (1-F_i(s_i))}{m\,(B_{i0} - E\,[B_i|s_i])\, f_i(s_i)}. \tag{C.5}$$

Then, by substituting Eqs. (C.3) and (C.4) into (C.1), the equation set, which would leads to the second best solution, is simplified to an equation about s_i, and shown as follows

$$(m(1-k_i c_i s_i) - (m+1)\boldsymbol{\alpha}_i) f_i(s_i)\boldsymbol{\rho}_i - N k_i c_i^2 + (1-F_i(s_i))\, k_i c_i \sum_{j=1}^{m} P_j N_j = 0. \tag{C.6}$$

That is to say, the problem is to find an optimal s_i to maximize the expected profit of customer j. In order to find the optimal solution of s_j, we first to prove that the existence of the solution.

The corresponding function of Eq. (C.6) can be written as follows:

$$g_i(s_i) = (m(1-k_i c_i s_i) - (m+1)\boldsymbol{\alpha}_i) f_i(s_i)\boldsymbol{\rho}_i - N k_i c_i^2 + (1-F_i(s_i))\, k_i c_i \sum_{j=1}^{m} P_j N_j.$$

where $\boldsymbol{\alpha}_i$ and $\boldsymbol{\rho}_i$ are given as Eqs. (C.4) and (C.5). From the approximation from the Poisson distribution, we have $F_i(0) \simeq 0$, and then there are $F_i(2\mu_i) \simeq 1, f_i(0) = f_i(2\mu_i) \simeq 0$, $E[B_i|s_i = 0] \simeq \mu_i$ and $E[B_i|s_i = 2\mu_i] \simeq 0$.

When $s_i = 0$, (i) If $B_{i0} > \mu_i$: from Eq. (C.4), we can know that $\boldsymbol{\alpha}_i \simeq 1 + \frac{Nc_i - \sum_{j=1}^m P_j N_j}{mNc_i(B_{i0}-\mu_i)f_i(0)} = -\infty$, and then from Eq. (C.5), there is $\boldsymbol{\rho}_i f_i(0) \simeq \frac{\sum_{j=1}^m P_j N_j - Nc_i}{m(B_{i0}-\mu_i)} > 0$. Consequently, $g_i(0) = +\infty$. (ii) If $B_{i0} < \mu_i$, similarly we can know that $\boldsymbol{\alpha}_i = +\infty$ and $\boldsymbol{\rho}_i f_i(0) \simeq \frac{\sum_{j=1}^m P_j N_j - Nc_i}{m(B_{i0}-\mu_i)} < 0$, and as a result, there is also $g_i(0) = +\infty$. In summary, $g_i(0) = +\infty$ exists.

When $s_i = 2\mu_i$, because $f_i(2\mu_i) \simeq 0$, $(1 - F_i(s_i))^2$ is an infinitesimal of higher order of $f_i(2\mu_i)$, that is, $(1 - F_i(s_i))^2 = \int_{2\mu_i}^{+\infty} f_i(x_i)dx = o\left(f_i(2\mu_i)\right)$. Consequently, according to Eq. (C.4), $\boldsymbol{\alpha}_i \simeq 1 + \frac{(1-F_i(s_i))^2}{mB_{i0}f_i(2\mu_i)} = 1^+$, further, from Eq. (C.5), $\boldsymbol{\rho}_i f_i(2\mu_i) \simeq \frac{-Nc_i(1-F_i(2\mu_i))}{mB_{i0}} = 0^-$. Accordingly, $g_i(2\mu_i) \simeq -\frac{k_i c_i^2}{P}$.

As a result, there must be at least one $s_i^* \in (0, 2\mu_i)$ which satisfied $g_i(s_i^*) = 0$. Suppose there are $r(\geq 1)$ $s_i^* \in (0, 2\mu_i)$, $s_{i1}^*, s_{i2}^*, \ldots, s_{ir}^*$, then we can find the optimal one through comparing the corresponding profits of the customers, that is,

$$s_i^{SB} = \arg_{s_{ik}^*} \left\{ \max \Pi_j \left(s_{ik}^* \right) \middle| k = 1, 2, \ldots, r \right\}.$$

Then, the optimal values of $\left\{ s_i^{SB}, \boldsymbol{\alpha}_i^{SB}, \boldsymbol{\rho}_i^{SB} \right\}$ can be obtained according to Eqs. (C.4) and (C.5). Furthermore, according to $\boldsymbol{\alpha}_i = \sum_{j=1}^m \alpha_{ij}$ and $\boldsymbol{\rho}_i = \sum_{j=1}^m \rho_{ij} P_j N_j$, the unique optimal equilibrium of the decentralized game can be obtained.

We then prove $\alpha_{ij}^{SB} = \alpha_{ik}^{SB}$ for $\forall j, \forall k = 1, 2, \ldots, m$. According to the analysis presented in Sect. 5.4, we know that the Nash equilibrium of the customers is determined by the following equations

$$1 - F_i(s_i^{SB}) = N \left(1 - \sum_{j=1}^m \alpha_{ij}^{SB} \right) \bigg/ \sum_{j=1}^m \rho_{ij}^{SB} P_j N_j, \tag{C.7}$$

$$\begin{aligned} \alpha_{ij}^{SB} = & \frac{\left(1 - \sum_{l=1}^m \alpha_{il}^{SB} - k_i c_i s_i^{SB}\right) f_i\left(s_i^{SB}\right) \sum_{l=1}^m \rho_{il}^{SB} P_l N_l}{f_i\left(s_i^{SB}\right) \sum_{l=1}^m \rho_{il}^{SB} P_l N_l + N k_i c_i^2} \\ & + \frac{\left(1 - \rho_{ij}^{SB}\right)\left(1 - F_i\left(s_i^{SB}\right)\right) k_i c_i P_j N_j}{f_i\left(s_i^{SB}\right) \sum_{l=1}^m \rho_{il}^{SB} P_l N_l + N k_i c_i^2}, \end{aligned} \tag{C.8}$$

$$\rho_{ij}^{SB} = 1 - \frac{N\alpha_{ij}^{SB} c_i \left(1 - F_i\left(s_i^{SB}\right)\right) + \left(B_{i0} - E[B_i|s_i^{SB}]\right) f_i\left(s_i^{SB}\right) \sum_{l=1}^m \rho_{il}^{SB} P_l N_l}{P_j N_j \left(1 - F_i\left(s_i^{SB}\right)\right)^2}. \tag{C.9}$$

For any given j and k, on the one hand, from Eq. (C.9), there is,

$$\begin{aligned}
&\rho_{ij}^{SB} P_j N_j - \rho_{ik}^{SB} P_k N_k \\
&\quad = \frac{(P_j N_j - P_k N_k)\left(1 - F_i\left(s_i^{SB}\right)\right)^2 - (\alpha_{ij}^{SB} - \alpha_{ik}^{SB})\left(1 - F_i\left(s_i^{SB}\right)\right) N c_i}{\left(1 - F_i\left(s_i^{SB}\right)\right)^2} \\
&\quad = (P_j N_j - P_k N_k) - (\alpha_{ij}^{SB} - \alpha_{ik}^{SB}) N c_i \big/ \left(1 - F_i\left(s_i^{SB}\right)\right).
\end{aligned} \tag{C.10}$$

On the other hand, from Eq. (C.8), we have

$$\alpha_{ij}^{SB} - \alpha_{ik}^{SB} = \frac{\left(1 - F_i\left(s_i^{SB}\right)\right) k_i c_i \left(P_j N_j - P_k N_k - \rho_{ij}^{SB} P_j N_j + \rho_{ik}^{SB} P_k N_k\right)}{f_i\left(s_i^{SB}\right) \sum_{l=1}^{m} \rho_{il}^{SB} P_l N_l + N k_i c_i^2}.$$

Then according to Eq. (C.10), there is

$$\alpha_{ij}^{SB} - \alpha_{ik}^{SB} = \frac{(\alpha_{ij}^{SB} - \alpha_{ik}^{SB}) N k_i c_i^2}{f_i\left(s_i^{SB}\right) \sum_{l=1}^{m} \rho_{il}^{SB} P_l N_l + N k_i c_i^2}.$$

Therefore, there is

$$\alpha_{ij}^{SB} - \alpha_{ik}^{SB} - \frac{(\alpha_{ij}^{SB} - \alpha_{ik}^{SB}) N k_i c_i^2}{f_i\left(s_i^{SB}\right) \sum_{l=1}^{m} \rho_{il}^{SB} P_l N_l + N k_i c_i^2} = 0,$$

and then

$$\frac{\left(\alpha_{ij}^{SB} - \alpha_{ik}^{SB}\right) f_i\left(s_i^{SB}\right) \sum_{l=1}^{m} \rho_{il}^{SB} P_l N_l}{f_i\left(s_i^{SB}\right) \sum_{l=1}^{m} \rho_{il}^{SB} P_l N_l + N k_i c_i^2} = 0.$$

However, from Eq. (C.7), we know that $f_i\left(s_i^{SB}\right) \sum_{l=1}^{m} \rho_{il}^{SB} P_l N_l$ cannot be zero. Consequently, $\alpha_{ij}^{SB} = \alpha_{ik}^{SB}$ exists.

MIX
Papier aus verantwortungsvollen Quellen
Paper from responsible sources
FSC® C105338

If you have any concerns about our products,
you can contact us on
ProductSafety@springernature.com

In case Publisher is established outside the EU,
the EU authorized representative is:
Springer Nature Customer Service Center GmbH
Europaplatz 3, 69115 Heidelberg, Germany

Printed by Libri Plureos GmbH
in Hamburg, Germany